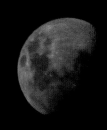

SPACE

A GUIDE TO THE INFINITE

CLIVE GIFFORD

President: Sean Moore
Production Director: Adam Moore
Editorial Director: Lisa Purcell
Editor: Finn Moore

Art Direction and Cover Design: Duncan Youel
at OilOften, London. www.oiloften.co.uk
Book Design: Nicola Plumb
Picture Research: Clive Gifford / Mike Plumb

ISBN: 978-1-62669-154-4

Printed and bound in China

10 9 8 7 6 5 4 3 2

SPACE

A GUIDE TO THE INFINITE

CLIVE GIFFORD

CONTENTS

The world may seem a big place, but astronomy tells us it is anything but. Earth is a wondrous yet small planet dwarfed by the unfathomably large scale that is all space and time contained within the Universe.

LOOKING UP AND OUT

As Kaitlin Casey, astronomer at the University of Texas at Austin, stated "It's rather humbling. Astronomy has taught us that we're not the centre of the Universe, we're not even at the centre of our Solar System or at the centre of our galaxy."

Whereas the Universe was once thought to revolve around Earth like a panoramic display, we now know that our planet plays by the gravitational rules of celestial mechanics. Earth humbly orbits a yellow dwarf star, the Sun, now

approaching the midpoint of its 10 billion year or so main sequence. Seven further planets orbited by more than 180 moons, five dwarf planets, thousands of asteroids and millions of comets all comprise the solar system that Earth is a part of. Yet the Sun is just one of between 200 and 400 billion stars found in the Milky Way, which itself is one of an estimated 200 billion to two trillion galaxies found throughout the Universe.

Human beings are an intensely curious species. Ever since the dawn of human

history, we have looked up and out at the night sky with wonder, awe, and an intense desire to know more. From the single viewpoint that Earth affords, talented scientists, astronomers, and cosmologists have made remarkable advances and discoveries to piece together a detailed body of knowledge about what the Universe is, what it contains, and how it works. Yet, like the Earth is dwarfed by the rest of the galaxy that it is a minuscule part of, what we know about the Universe may just be the beginning.

The outstanding beauty and clarity of the
Milky Way above the Utah desert, USA.
LEFT: Milky way over The Church of Good
Shepherd, Lake Tekapo, New Zealand

7

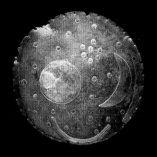

No one knows when or where people first started to do more than idly gaze upwards at night, but started to track and note the movement of stars, planets and galaxies across the sky.

ANCIENT STARGAZING

ABOVE: The Nebra sky disk, discovered in 1999, is one of the oldest known representations of the Universe.
LEFT: A representation of the Mayan calendar system carved in stone. The Mayans used three different calendars lasting 260 days, 365 days, and over 7,800 years. / A 15th Century Equatorial Armilla in Beijing, China, used to measure the celestial latitude and ascension of objects in the night sky.

Ancient peoples associated stars with their deities and religious beliefs. They also used their observations for more practical purposes such as navigation, timekeeping, marking the passing of the seasons, and devising lunar and solar calendars. Observations of movements across the night sky were part astronomy, part astrology or divination, and heavily linked with and cloaked in religious belief.

Many ceremonial structures, from standing stones to complete temples, were orientated in such a way that they interacted or aligned with the Sun, Moon, or stars' position in the sky. The Intihuatana stone at the Inca city of Machu Picchu, for example, is perfectly aligned with the Sun's position during the winter solstice. The pre-Columbian city of Chichen Itza features giant temples that act as solar calendars, as well as El Caracol (the Observatory) whose windows and doors are aligned to the movement of the Sun, Venus and other planets.

One of the earliest known astronomical artifacts is the Nebra sky disk, which was unearthed in Saxony-Anhalt, Germany and dates to about or before 1600BCE. This bronze circle depicts the Sun, lunar crescent and a number of stars including the Pleiades open star cluster. The same cluster was plotted and its position was derived by the ancient Greek astronomer, Thales of Miletus. According to the historian Plutarch, Thales was also the first to plot the Hyades cluster and to predict the solar eclipse of 585BCE.

He was followed by a sequence of enterprising classical world astronomers who helped turned stargazing into something closer to a rigorous science. Star catalogues and constellation guides were developed in classical ancient world cultures around the Mediterranean, in the Middle East, and in China and Korea. By around 4BCE, for example, the Chinese astronomer Shi Shen had catalogued more than 800 stars and 122 constellations. Claudius Ptolemy, another star cataloguer based in the Egyptian city of Alexandria, produced a highly influential description of the Solar System around 150CE. His thirteen-volume work, The Almagest, subscribed to a geocentric view of space with Earth at its centre that would be held as factually correct for more than a millennium.

THIS PAGE: Intihuatana, an astronomical calendar device known as the "hitching post of the Sun," is found in the Inca city of Machu Picchu, Peru.

Around 270BCE, an ancient Greek named Aristarchus of Samos proposed an alternative Solar System model to geocentrism, placing the Sun at the center with Earth and the other known planets orbiting about it on circular paths.

ANCIENT STARGAZING

LEFT: Sir Isaac Newton (1642-1727) examining the nature of light, splitting white light into a spectrum of colours with the aid of a crystal prism.
ABOVE: Hipparchus observing celestial bodies at the Observatory of Alexandria, Egypt. This second century BCE astronomer compiled the Western world's first major star catalogue comprising at least 850 entries.

This heliocentric model would take over 1,800 years to take hold in Europe. As the Ask an Astronomer website so eloquently puts it: "Europe muddled through the Dark Ages," while "astronomers in the Middle East translated Greek texts into Arabic, preserving and expanding humanity's knowledge of the sky." A handful of Arabic astronomers challenged Ptolemy's geocentric model before Aristarchus' concept was fully revisited by Polish mathematician and astronomer Nicolaus Copernicus. His De Revolutionibus Orbium Coelestium (On the Revolutions of the Heavenly Spheres) was completed in 1532 but wasn't published for eleven years, just several months before Copernicus' death, for fear of incurring the wrath of the Catholic Church who strongly supported the geocentric model.

Copernicus's theories proved influential, especially in the 17th Century when astronomy took a number of giant leaps forward, aided by the invention of the optical refracting telescope in the first decade and the work of such luminaries as Johannes Kepler (devisor of the three laws of planetary motion and describing orbiting bodies' paths as elliptical rather than circular), Christiaan Huygens and Isaac Newton. Newton is believed to have developed the first reflecting telescope eschewing glass lenses in favour of mirrors and in 1687 published his ground-breaking Philosophiae Naturalis Principia Mathematica, establishing his theory of universal gravitation and the laws of motion which would govern people's investigations from then on.

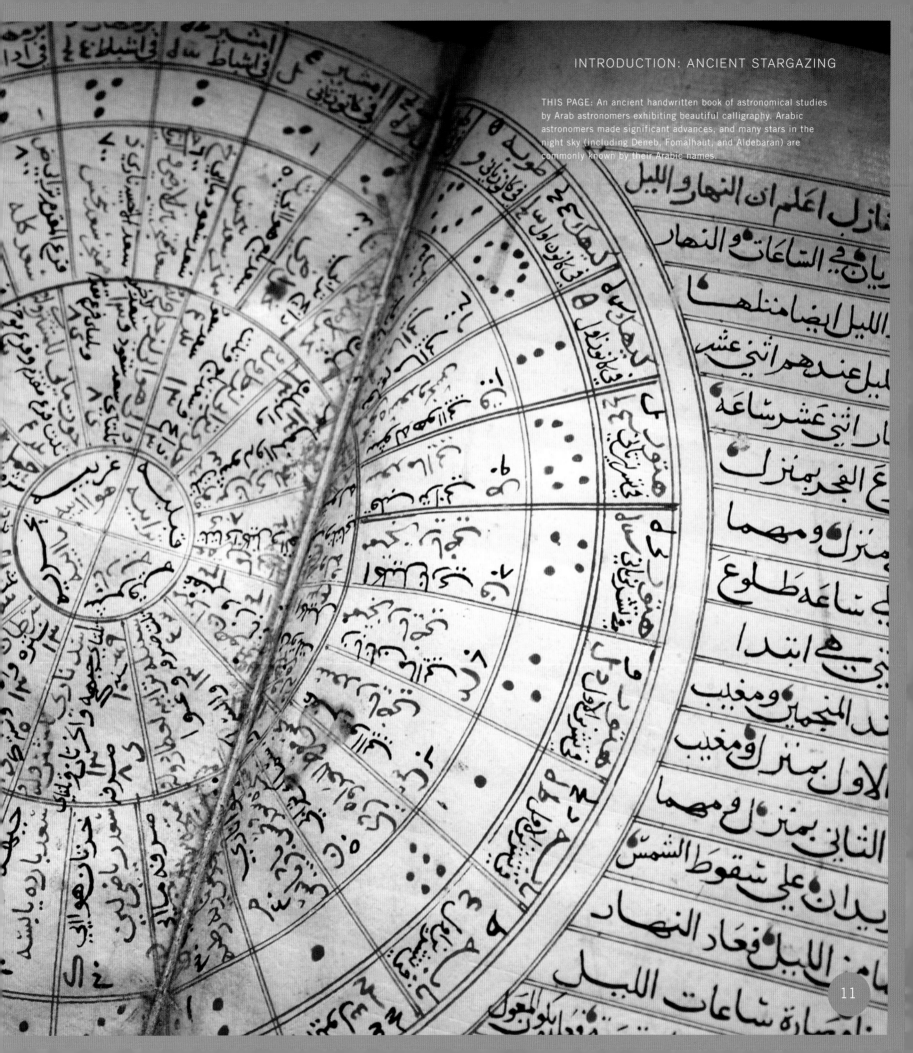

THIS PAGE: An ancient handwritten book of astronomical studies
by Arab astronomers exhibiting beautiful calligraphy. Arabic
astronomers made significant advances, and many stars in the
night sky (including Deneb, Fomalhaut, and Aldebaran) are
commonly known by their Arabic names.

The Big Bang – the theory that the Universe had inflated and expanded out of a single point – was gaining hold by the 1960s but the work of two radio astronomers would help underpin its veracity by providing solid evidence.

ECHOES OF CREATION

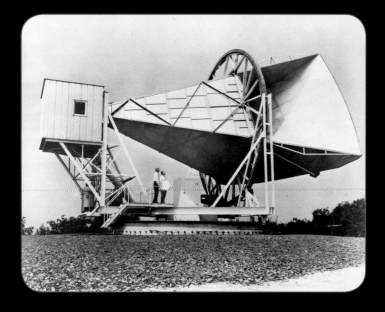

LEFT: The Holmdel horn antenna at Bell Telephone Laboratories in Holmdel, New Jersey. Built in 1959 to work with NASA's pioneering ECHO communication satellites, the antenna measured 50 feet in length with a radiating aperture of 20 feet by 20 feet. The entire structure was composed of aluminum with a base of steel and weighed approximately 18 tons.

Robert Wilson and Arno Penzias were working with the 50 foot long, 18 ton Holmdel horn antenna at Bell Telephone Laboratories in New Jersey in 1964 and 1965 mapping out microwave signals from the Milky Way. The pair found that they couldn't remove a slight background noise that appeared to emanate from all directions in space.

They checked the horn antenna's connections, wiring and even ensured the interior was free of pigeons and their droppings but still the background hum remained. As Penzias later said, "We didn't understand its significance, and we never dreamed it would be connected to the origins of the universe."

What the pair had detected was background microwave radiation left over some 380,000 years after the Big Bang – the earliest radiation yet detected. Predictions had been made that if the Big Bang had occurred, then low level radiation would still exist and be found throughout the Universe. Penzias and Wilson's discovery – now called the Cosmic Microwave Background – was it. Successive satellites including Cosmic Background Explorer (COBE) and the Wilkinson Microwave Anisotropies Probe (WMAP) have investigated it further to find clues and insight into the formation of the Universe. Wilson and Penzias were awarded the 1978 Nobel Prize for physics for their discovery (shared with the Soviet physicist, Pyotr Kapitsa).

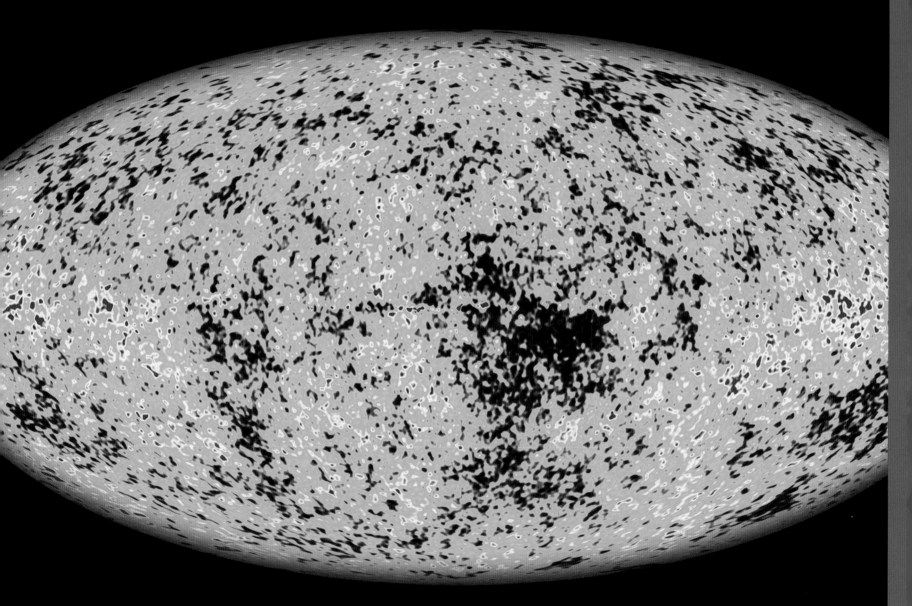

THIS PAGE: An image of the CMB (cosmic
microwave background radiation) taken by the
WMAP (Wilkinson Microwave Anisotropy Probe)
- a space-based microwave telescope, launched
in 2001 which measured temperature differences
across the sky in the CMB.

THE
SOLAR SYSTEM

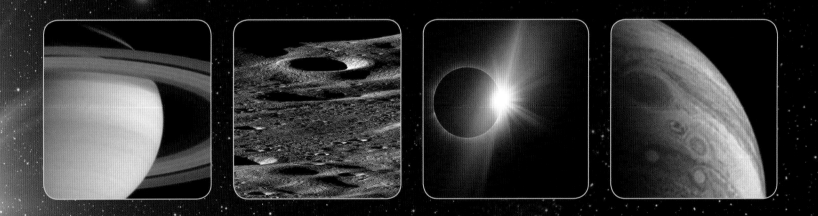

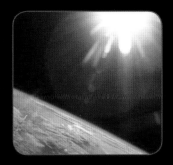

This 865,374 mile wide ball of flaming gases is responsible for both heating and illuminating our planet.

THE SUN

The Sun is a yellow dwarf star, over 109 times wider than Earth and with approximately 1.3 million times more volume than Earth. It comprises over 99.8 per cent of the entire mass of the Solar System, equivalent to more than 330,000 Earth masses. With such mass comes a hefty gravitational pull that keeps asteroids, eight planets and distant comets on the far reaches of the Solar System within its orbit. The Sun rotates about its own vertical axis, but since it is not a solid body, different parts rotate at different rates – from 36 days at the Sun's poles to every 25 days at the solar equator.

Powering up into its main sequence more than 4.5 billion years ago, nuclear fusion reactions at the Sun's core result in temperatures at its center reaching 27 million degrees Fahrenheit as fusion of hydrogen produces vast amounts of energy. Newly-generated energy can take 100,000 years to rise up through the radiative and convective zones of the Sun (the convective zone is estimated as spanning 124,275 miles in depth) until the energy reaches the Sun's surface or photosphere which has a surface temperature of 9,930 degrees Fahrenheit. That may sound surprisingly cool in comparison to the core, but as NASA

pointedly notes, "it's still enough to make carbon, like diamonds and graphite, not just melt, but boil." When emitted from the surface, the Sun's energy travels at the speed of light to reach Earth in a little over eight minutes. The Sun's atmosphere is divided into an inner zone or chromosphere and outer atmosphere or corona which extends many millions of miles into space.

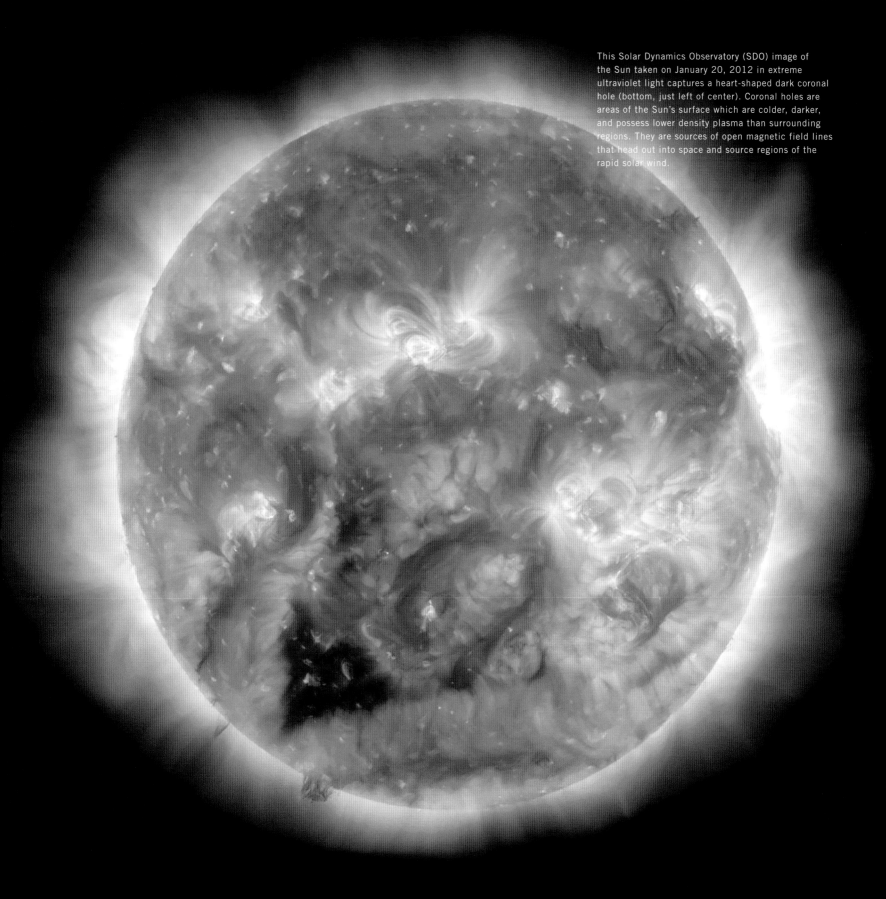

This Solar Dynamics Observatory (SDO) image of the Sun taken on January 20, 2012 in extreme ultraviolet light captures a heart-shaped dark coronal hole (bottom, just left of center). Coronal holes are areas of the Sun's surface which are colder, darker, and possess lower density plasma than surrounding regions. They are sources of open magnetic field lines that head out into space and source regions of the rapid solar wind.

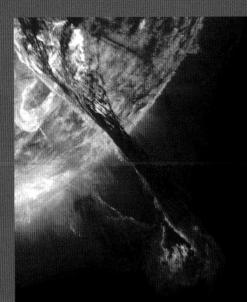

A close inspection of the Sun's photosphere shows all is far from serene on its turbulent visible surface.

SUNSPOTS, PROMINENCES AND SOLAR FLARES

A complex magnetic field generated by the Sun's electric currents extends out far into space (the heliosphere) and is carried by charged particles as the solar wind. Dark patches or depressions located on the Sun's surface and known as sunspots are significantly cooler than the surrounding areas, sometimes by as much as 3,600 degrees Fahrenheit. They can emerge and last for anywhere between a few hours and several months and are typically in the range of 900 to 31,000 miles in width. Depth measurements can be hard to gauge but solar astronomers have observed some sunspots with depths of more than 800 miles. Sunspots increase and decrease in number on an approximately 11 year cycle with some variation and were first measured and recorded in 1749.

Sunspots tend to accompany violent activity on the solar surface including flaming jets and loops of hot gas and plasma called prominences. On August 5th, 2012, one of the largest known prominences was observed - a filament stretching almost 500,000 miles across the face of the Sun, more than a solar radius in length. Coronal Mass Ejections and solar flares are large explosions on the photosphere capable of generating heat measured in millions of degrees Fahrenheit and, according to the NOAA's National Weather Service, releasing as much energy as a billion megatons of TNT explosive. Interactions between hot plasma and the Sun's magnetic field hurl a burst of plasma up and away from the Sun in the form of a solar flare emitting large quantities of electromagnetic radiation.

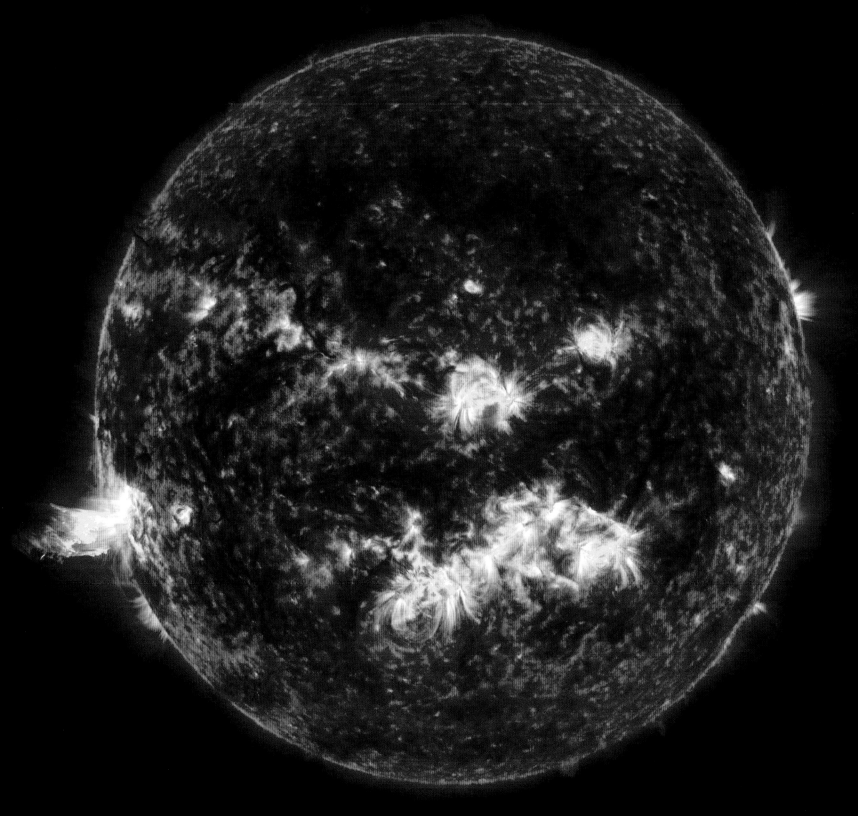

THIS PAGE: A solar flare and eruption of solar material shooting through the Sun's atmosphere at high speed in June 2013. Known as a prominence eruption, this event was followed by a coronal mass ejection out into space.

LEFT: A coronal mass ejection (CME) observed in the summer of 2012. Long filaments of solar material erupted out into space, travelling at a velocity of more than 900 miles per second. The ejection interacted with Earth's magnetosphere, prompting beautiful aurora displays in Earth's atmospheres.

An eclipse is the obscuring of the light from one celestial body by the passage of another between it and the observer, or between it and its source of illumination.

LUNAR AND SOLAR ECLIPSES

SEQUENCE OF THE LUNAR ECLIPSE OF FEBRUARY 20, 2008

Lunar eclipses occur when the Earth blocks the sunlight that is normally reflected by the Moon. Instead, our satellite is bathed in Earth's shadow plus a fraction of its normal quota of sunlight that passes through Earth's atmosphere where much of the blue light is filtered out and only longer, red wavelengths of light reach the Moon. This light gives our satellite an orange or red appearance from our viewpoint on Earth, inspiring ancient myths such as the Inca stories of a jaguar bloodying the Moon.

A total lunar eclipse will last for less than two hours and is safe to view without eye protection. Whilst the impact on Earth is minimal, the effects on the Moon can be striking. Instruments at the Apollo 12 landing site noted a large drop a 321.3 degrees Fahrenheit drop in temperature during a 1971 total eclipse.

A solar eclipse sees the Moon play the blocking role, only partially blocking out the Sun on most occasions with total solar eclipses rarer and only viewable by a portion of the planet. During the brief period of totality, when the Sun is completely covered in a total solar eclipse, the Sun's outer atmosphere, the corona, normally hard or impossible to view, is revealed. Totality is fleeting, lasting a maximum of seven and a half minutes although most total eclipses are much shorter.

Solar Eclipse (Elements of this image furnished by NASA).

This composite image of seven pictures shows the progression of a partial solar eclipse from Ross Lake, Northern Cascades National Park, Washington on August 21, 2017. The second to the last frame shows the International Space Station, with a crew of six onboard, in silhouette as it transits the Sun at roughly five miles per second.

A total solar eclipse swept across a narrow portion of the contiguous United States from Lincoln Beach, Oregon to Charleston, South Carolina. A partial solar eclipse was visible across the entire North American continent along with parts of South America, Africa, and Europe.

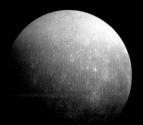

The planet nearest the Sun is also the smallest, with a diameter of 3,032 miles. It is only 38 per cent the diameter of Earth and has just one fifteenth of our planet's surface area.

MERCURY

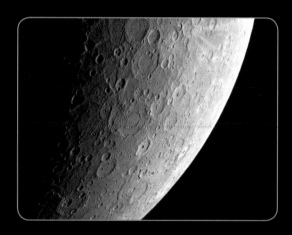

ABOVE: The smallest planet also has the smallest tilt of any Solar System planet. The planet's volume is just 0.056 that of Earth but has a similar, although slightly lesser, density to Earth.

LEFT: Taken by the MESSENGER spacecraft's Wide Angle Camera (WAC), this image of Mercury shows a portion of the planet's heavily cratered surface. Mercury's mantle and crust combined are around 250 miles in thickness and surround a heavy, metallic core, some of which planetary scientists believe may be liquid. The planet's magnetic field is only one per cent that of Earth's.

Mercury's close proximity to the Sun (as close as 28.6 million miles at its perigee with our star) means the Sun would loom more than three times as large if viewed from the planet's surface. It also causes Mercury, named after the Roman messenger god, to race through its eccentric, egg-shaped orbit with uncommon haste, travelling at a velocity of 105,840 mph – the fastest-moving of all the planets. Mercury rotates very slowly. This means that each day on the planet lasts 175.9 Earth days making a day on

Mercury longer than its year. Brutally scarred by thousands of craters where asteroids and meteorites have bombarded its surface and left their telling mark, the planet's pock-marked surface includes its deepest point (the 17,650-ft deep Rachmaninoff crater) and its largest crater, the Caloris Basin which dominates the landscape of the planet's northern hemisphere. At 947.6 miles in diameter, the crater formed by impact over 3.8 billion years ago, would extend farther than London to

Rome. Away from the craters, long cliffs called scarps extend many hundreds of kilometres and rise up to 1 mile high. Mercury's average surface temperature is a blistering 332.6 degrees Fahrenheit, but this bald figure doesn't tell the whole story. With little atmosphere to protect its battleworn surface, Mercury experiences wild temperature swings; at its hottest in the Sun's full glare, temperatures on its surface can reach 800 degrees Fahrenheit.

Mercury. It is the smallest and closest to the
Sun of the eight planets in the Solar System,
with an orbital period of about 88 Earth days.
Elements of this image furnished by NASA.

Lying an average of 67.24 million miles from the Sun, Venus is Earth's nearest planetary neighbour and often described as its twin due to the great similarities in the two planets' diameter, mass, and structure with both, possessing an iron core, rocky mantle, and solid crust.

VENUS

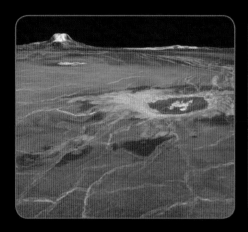

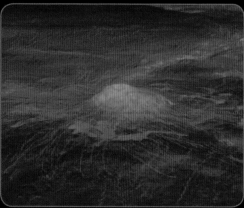

ABOVE: Hemispheric view of Venus, centered at a longitude of 270 degree east and taken by combining large amounts of radar data derived from NASA's successful Magellan mission.

FAR LEFT: A portion of western Eistla Regio – a 5000 mile long stretch of the Venusian surface is displayed in this three-dimensional perspective built up from imagery from NASA's Magellan probe.

LEFT: This view of the volcanic peak Idunn Mons in the Imdr Regio area of Venus was formed from data gathered by both Magellan and the European Space Agency's Venus Express spacecraft which orbited the planet from 2006 to 2015. Planetary scientists speculate that this volcano is currently active.

Venus has approximately 91 per cent of the gravity experienced on Earth and possesses the sorts of terrain features familiar to us, including valleys, highland areas, and mountains, the highest of which (Maxwell Montes) rises almost 6.8 miles above the planetary mean elevation.

The Venusian surface is dominated by large volcanic plains of basalt called planitiae crossed with numerous channels called valles. Dali Chasma is an intricate network of troughs and canyons that run

a total length of 1291 miles. There are two large highland areas. In the north polar region lies Ishtar Terra, roughly the size of Australia, whilst to the south and crossing the planet's equator is Aphrodite Terra, a region of uplands similar in size to the South American continent.

More than 1,600 major volcanoes have been identified on the planet's surface with the possibility of tens of thousands more. Many are shield volcanoes formed from extremely thick, viscous lavas.

Pancake domes – a volcanic feature unique to the planet – are low, flat-topped domes that can run 40 miles or more wide but with little elevation. According to NASA, it is thought that Venus was completely resurfaced by volcanic activity 300 to 500 million years ago. The impact craters that exist on its surface therefore are all young and range from several miles in width up to Mead Crater which spans 168 miles across.

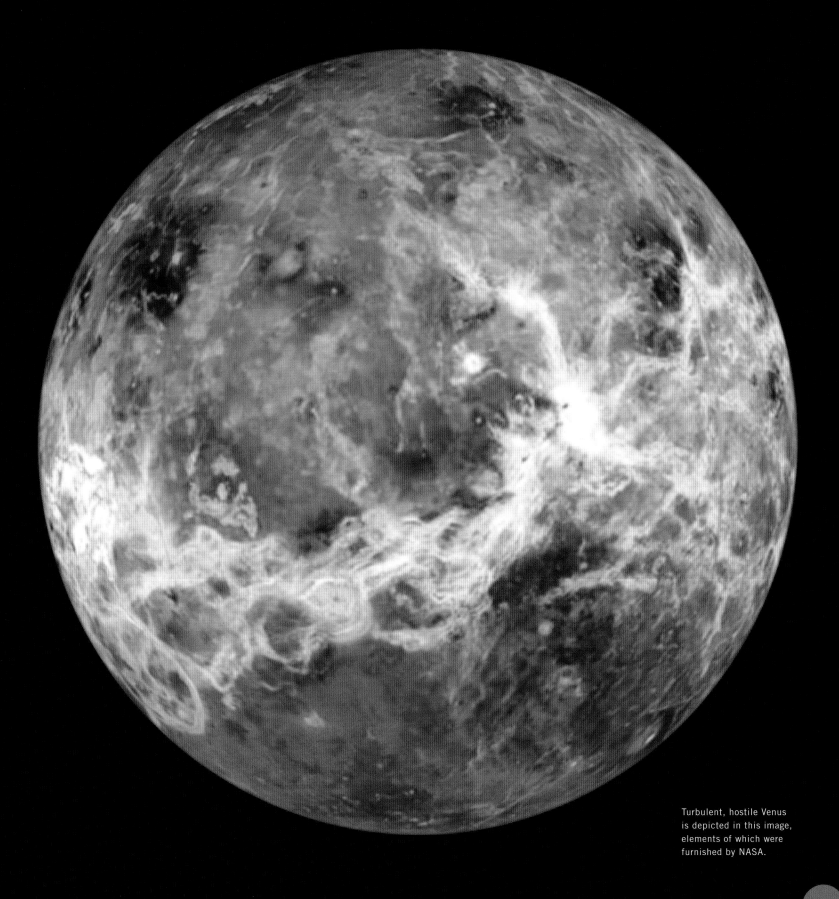

Turbulent, hostile Venus is depicted in this image, elements of which were furnished by NASA.

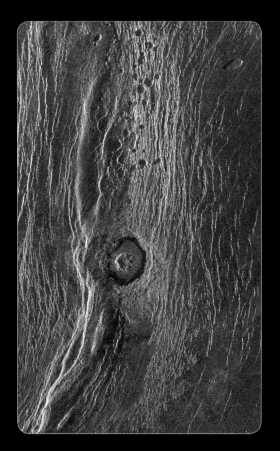

ABOVE: The northern part of the Akna Montes mountain range of Venus. This 520-mile-long range extends north to south and surrounded by the deformed plains of the plateau, Lakshmi Planum.

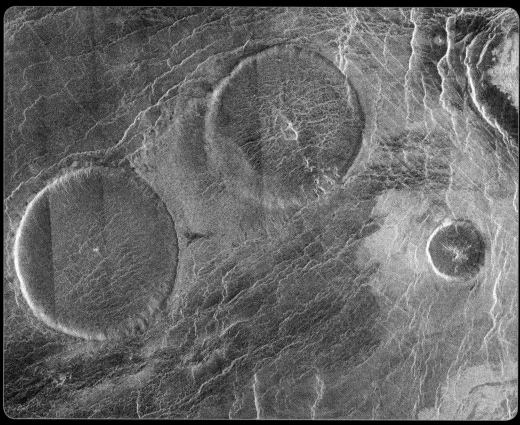

ABOVE: The unusual pancake-shaped domes found on Tinatin Planitia – a low plain named after a heroine in Georgian mythology. The domes have been formed by extruding lava of high viscosity which explains the dome's steep edges and flattened tops. The largest dome measures just under 39 miles in diameter.

Venus rotates about its own axis with a miniscule tilt of just three degrees. It turns in a retrograde motion and slowly, taking 243 days to complete a revolution, 18 days more than it takes to complete its orbit around the Sun. The planet is blanketed in a thick, suffocating atmosphere where carbon dioxide, not nitrogen or oxygen, is the prime constituent. CO_2 comprises almost 96.5 per cent of the entire atmosphere, which as a whole presses down on the surface of the planet with 90 times more pressure than experienced on Earth – the equivalent of being around 3,000 feet underwater.

The dense atmosphere rises from surface level up beyond 60 miles and reflects four-fifths of the incoming solar radiation but traps most of the heat that does enter the atmosphere or strikes the surface. This results in a runaway greenhouse effect with surface temperatures of 880 degrees Fahrenheit. Temperatures vary amongst the layers of dense clouds with the top top layer propelled by hurricane-force winds travelling up to 225 mph and circulating the planet once every four Earth days. Large opaque clouds of highly acidic sulphur dioxide merely add to the hostility of the planet's atmospheric conditions.

As the closest planet to us, with its surface obscured from view to optical telescopes and with mysteries abounding, it is no surprise that Venus has been the target of many missions, the first in 1962 when Mariner 2 achieved a distant fly-by of the planet. The Soviet Union Venera series of probes achieved a number of milestones including Venera 7 which in 1970 became the first probe to touch down on and send data back from another planet's surface. Two decades later, NASA's Magellan spacecraft achieved orbit insertion around Venus and spent the next four years mapping the planet's surface in unprecedented detail using Synthetic Aperture Radar. The European Space Agency's Venus Express craft went into polar orbit around the planet in April 2006 and operated for eight and a half years, discovering a thin ozone layer and an ultra-cold layer within the planet's atmosphere as well as a huge double atmospheric vortex at the south pole.

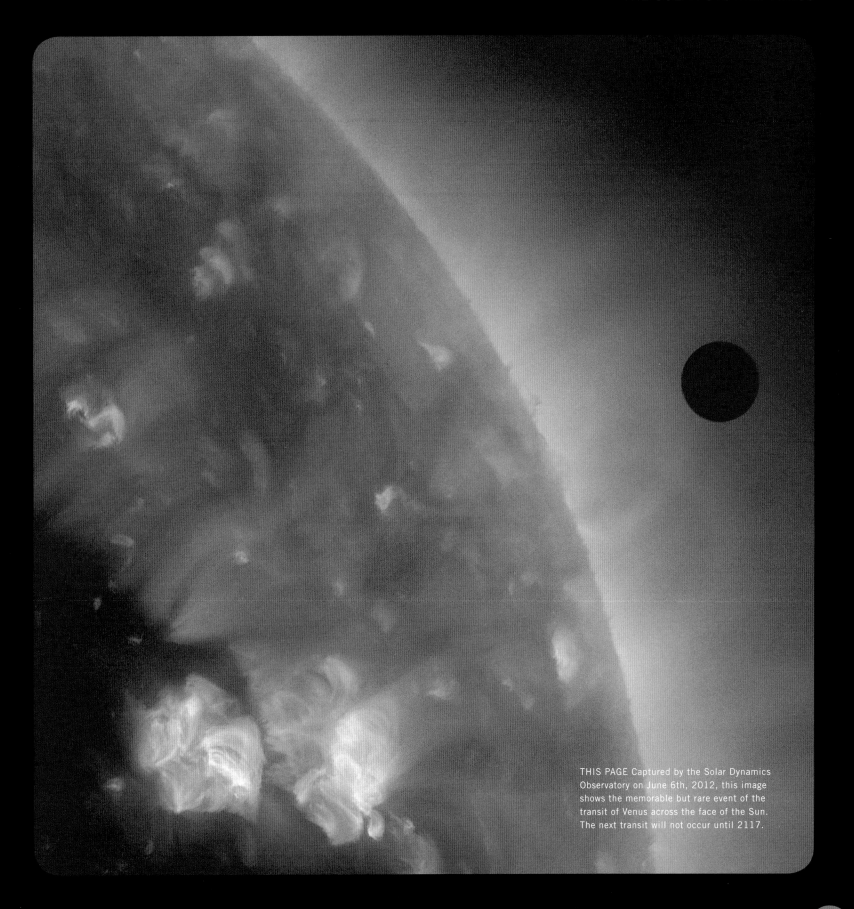

THIS PAGE Captured by the Solar Dynamics Observatory on June 6th, 2012, this image shows the memorable but rare event of the transit of Venus across the face of the Sun. The next transit will not occur until 2117.

Earth's sole natural satellite, the Moon, has fascinated and formed the basis of calendars, ceremonies and timekeeping over millennia as people watched it go through its phases, waxing and waning over its mean 29.5-day cycle.

THE MOON

ABOVE: Selected phases of the Moon (from left to right): waxing crescent, waxing quarter, waxing gibbous, full Moon, waning gibbous, waning quarter and waning crescent.

This rocky body has a core, mantle and crust and is believed to have been formed after a Mars-sized object collided with Earth over four billion years ago. It has been Earth's constant companion ever since, helping to moderate any wobble of our planet's axis and through mutual gravitational attraction with the Earth, causing the bulge in the world's waters known as ocean tides and in return pushing the Moon away fractionally each year an additional inch and a half.

Locked into a synchronous rotational mode with Earth, the Moon's rotation is such that it always presents the same side or face to us as it completes its orbit 238,855 miles away. The dark side of the Moon was much speculated about and even feared in the past until the Soviet Luna 3 probe in 1959 and subsequent probes viewed its rugged, heavily cratered terrain. The first humans to set eyes upon its surface were Frank F. Borman II, James A. Lovell Jr and William Anders on board Apollo 8 in 1968. Fifty-one years later, the Chinese lunar probe, Chang'e-4, landed and transmitted first-hand images of the dark side's surface.

Less dense and far smaller than Earth with an equatorial diameter of 2,160 miles, the Moon has 81 times less mass than Earth and its gravity is approximately one-sixth of what its parent planet experiences. Just four gases: argon, helium, hydrogen and its most abundant gas, neon, comprise 98 per cent of its tenuous atmosphere estimated to have a total mass of 55,000 pounds – a load that would not trouble a large dump truck.

This image of a full Moon with the Tycho crater particularly prominent (center, near bottom). The crater is named after renowned Danish medieval astronomer, Tycho Brahe, and is estimated at 108 million years of age.

When Italian scientist Galileo Galilei became the first person to train a telescope upwards to examine the Moon in November 1609, he was astonished to discover a far from perfect sphere but a surface comprising plains, depressions, craters and mountains.

THE MOON'S SURFACE

LEFT: This view of the north polar region of the Moon was obtained by NASA's Galileo spacecraft during its fly-by on December 7 and 8, 1992.

The Moon features a number of mountain ranges (called Montes) including Montes Taurus, Jura, and Caucasus. They vary in length fom the 30 mile long Montes Secchi to the 490 miles of Montes Rook and the 370 mile long Montes Apenninus, named after the Pennines in England, and containing the Moon's tallest mountain, Mons Huygens with an elevation in excess of 17,500 feet.

Lacking both tectonic plates and volcanic action, the vast majority of the Moon's mountainous features are considered the work of impacts in the past. The same is

the case with the vast number of craters that pock mark the lunar surface – a legacy of lacking a dense atmosphere to impede the progress of incoming bodies from space. German astronomer Johann Hieronymus Schröter was the first to name craters from the Greek krater for a large bowl or container in 1791. Total crater estimates vary wildly but could be in their millions. Scientific focus has been on larger craters. In 2010, for example, the LOLA instrument on board the Lunar Reconnaissance Orbiter measured over 5,000 craters above 12.5 miles in diameter. The largest, the South Pole–

Aitken basin, measures 1,600 miles in diameter and 8.1 miles deep.

More than a seventh of the Moon's surface is made up of flattish plains, called maria (singularly, mare) after the Latin for sea. Dry and often dusty, parts of these plains are covered in rocky rubble and peppered with craters. Mare Tranquillitatis or the Sea of Tranquility remains in the minds of many as the historic landing site of the Apollo 11 mission which placed two men, Neil Armstrong and Edwin 'Buzz' Aldrin on the lunar surface for the first time.

ABOVE: The barren, rocky landscape of the Moon boasts
features mostly formed from the results of impacts
as seen by the ongoing Lunar Reconnaissance Orbiter
mission which launched in 2009.

On July 21, 1969, 109 hours and 22.25 minutes after launch, Neil Armstrong emerged from his Apollo 11 Lunar Module *Eagle*, clambered down its external ladder and, set foot on the Moon.

MEN ON THE MOON - ONE SMALL STEP...

FAR LEFT: The iconic image of Edwin E. Aldrin Jr. on the lunar surface was taken by fellow Apollo 11 astronaut, Neil Armstrong during their historic EVA on the Moon. During their two and a half hour-long EVA, the pair planted a US flag, deployed a solar wind composition experiment amongst other equipment packages and collected some 46 pounds of soil and rock samples.

LEFT: The Saturn V launch vehicle lifts off from the Kennedy Space Center (KSC), Florida on July 16th, 1969 carrying the Apollo 11 spacecraft and crew. The triumphant three man Apollo crew returned to Earth on July 24th, splashing down in the Pacific Ocean.

An estimated one-seventh of the entire global population watched his every move more than 240,000 miles away. Within 20 minutes, Armstrong was joined by fellow crew member Edwin 'Buzz' Aldrin and the pair would spent just under 22 hours on the Moon's surface, mostly inside the Lunar Module, before ascending, re-docking with the Command and Service Module (CSM) containing Michael Collins in a lunar orbit and returning to Earth. The crew's completion of their historic 953,054-mile mission was the culmination of extraordinary political will, technical innovation, large-scale organization and

vision as America found itself locked in a 'Space Race' with rival superpower, the Soviet Union for supremacy.

The Apollo program consumed vast resources. According to Fraunhofer IEE Executive Director, Professor Clemens Hoffman, the Apollo project cost in excess of 25 billion dollars and at its peak in 1966, NASA occupied 4.4% of the entire US federal government's budget. A network of some 34,000 NASA employees and 375,000 outside contractors, scientists, engineers and technicians all contributed to the project – a reflection

of the vast number of systems and sub-systems required to send humans to the Moon. Five further Apollo crews would land safely on the lunar surface, the last three (Apollo 15, 16 and 17) deploying lunar rovers which extended the astronauts' range of exploration. Whilst the Apollo missions left behind rovers, descent modules, plaques, flags, and unwanted debris including urine bags, the Apollo crews returned a total of 842 pounds of lunar rocks and soil for investigation back to Earth.

"Now is the time...for a great new American enterprise...I believe that this nation should commit itself to achieving the goal, before this decade is out, of landing a man on the Moon and returning him safely to the Earth." President John F. Kennedy, May 1961.

BELOW: Apollo 11 astronaut, Edwin E. Aldrin Jr. stands close to the Stars and Stripes planted in the Sea of Tranquility region of the Moon in 1969. The store-bought flag cost just $5.50 but was modified due to the Moon's tenuous atmosphere and lack of breeze, by sewing a crossbar into a hem along the top of the flag for it to be clearly displayed to the more than 500 million TV viewers back on Earth.

A supermoon is the coincidence of a full moon or a new moon with the closest approach the Moon makes to the Earth on its elliptical orbit, resulting in the largest apparent size of the lunar disk as seen from Earth.

SUPERMOONS

LEFT: A supermoon rising in the Negev desert, Israel.

RIGHT: An impossibly large supermoon rises over the Pacific Ocean in this artist's impression. In reality, supermoons may appear fractionally larger than our usual view of the Moon as they rise above the horizon.

Looming a little larger in the sky than usual, supermoons excite public interest and encourage people to step outside and commune with the night sky. They occur when a full or new moon is at its closest to Earth. The Moon's orbit isn't perfectly symmetrical. Like many bodies, it has an apogee (its furthest point away) and perigee (its nearest point). In the Moon's case these mean the Moon can vary on average between 251,900 miles and 225,300 miles from Earth. When a new moon or full moon coincides with a perigee, a supermoon occurs.

The term 'supermoon' is not astronomical but one derived from the world of astrology. It was coined by the astrologer Richard Nolle in 1979 after he had read a paper on spring tides written by National Oceanic and Atmospheric Administration (NOAA) scientist, Fergus Wood. Many astronomers prefer the term perigee-syzygy to describe the event. The shape of the Moon's orbit changes over time due to the gravitational influence of the Sun and the other planets. This can mean that, occasionally, the Moon's perigee can be smaller than average, down to around

221,000 miles, thus increasing its size and brightness in the night sky. There were three supermoons in 2019 and four in 2020. The full moon of January 20-21, 2019 received plenty of media attention as it coincided with a total lunar eclipse. The Moon appeared to shine a rich red color from light reflected from Earth, a phenomenon known as a blood moon.

The third rocky planet from the Sun is the only
known place in the Universe to support life.

EARTH

The third rocky planet from the Sun is
the only known place in the Universe to
support life. Its oxygen-rich atmosphere,
abundance of surface water, crucial
nutrient and water cycles as well as
mild temperatures have enabled an
extraordinary biodiversity with over 1.8
million identified different species of
living thing from microscopic viruses and
single-celled organisms to blue whales, up
to 100 feet long and weighing in excess
of 130 tons. Trillions of living things,
including 7.8 billion humans, all co-exist
on the Solar System's fifth largest planet.

With an equatorial diameter of 7917.6
miles as measured by NASA, Earth is an
oblate spheroid with a slight belly bulge,
making it 26.6 miles wider than it is tall.
It boasts a surface area of 197 million
square miles. As it travels around the
Sun at an orbital velocity of 66,622mph,
it remains tilted at an angle of 23.4°

relative to the Sun's surface – a condition
which helps create the distinct seasons
that many parts of the planet experience.
Its orbital period (365.25 days) and
rotational period (23.9 hours) provide us
with our year and days. Whilst the orbit
varies, the planet's average distance from
the Sun of 92,956,050 miles puts it into
the habitable or "Goldilocks Zone," where
climatic conditions are neither too hot
nor too cold, but provide an environment
suitable for life to flourish. Life on Earth
is thought to have begun within the
planet's first billion years. It may have
originated as far back as 4.1 billion
years ago after the recent discovery of
potential fossilised bacteria in Quebec,
Canada reported in the journal, Nature.
What is certain is that life underwent vast
changes and endured massive climate
change and mass extinctions to reach the
point it has today.

BELOW: This composite image of southern Africa and the surrounding oceans was captured by six orbits of the NASA/NOAA Suomi National Polar-orbiting Partnership spacecraft on April 9, 2015, by the Visible Infrared Imaging Radiometer Suite (VIIRS) instrument. Tropical Cyclone Joalane can be seen over the Indian Ocean. Winds, tides and density differences constantly stir the oceans while phytoplankton continually grow and die. Orbiting radiometers such as VIIRS allows scientists to track this variability over time and contribute to better understanding of ocean processes that are beneficial to human survival on Earth. The image was created by the Ocean Biology Processing Group at NASA's Goddard Space Flight Center in Greenbelt, Maryland.

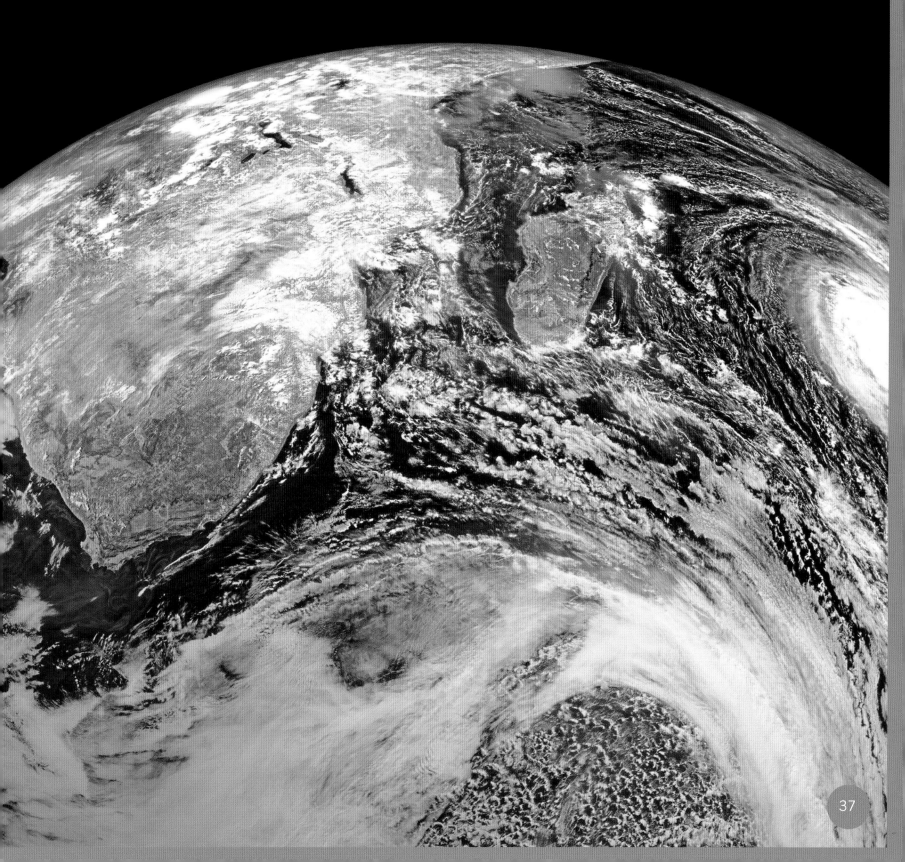

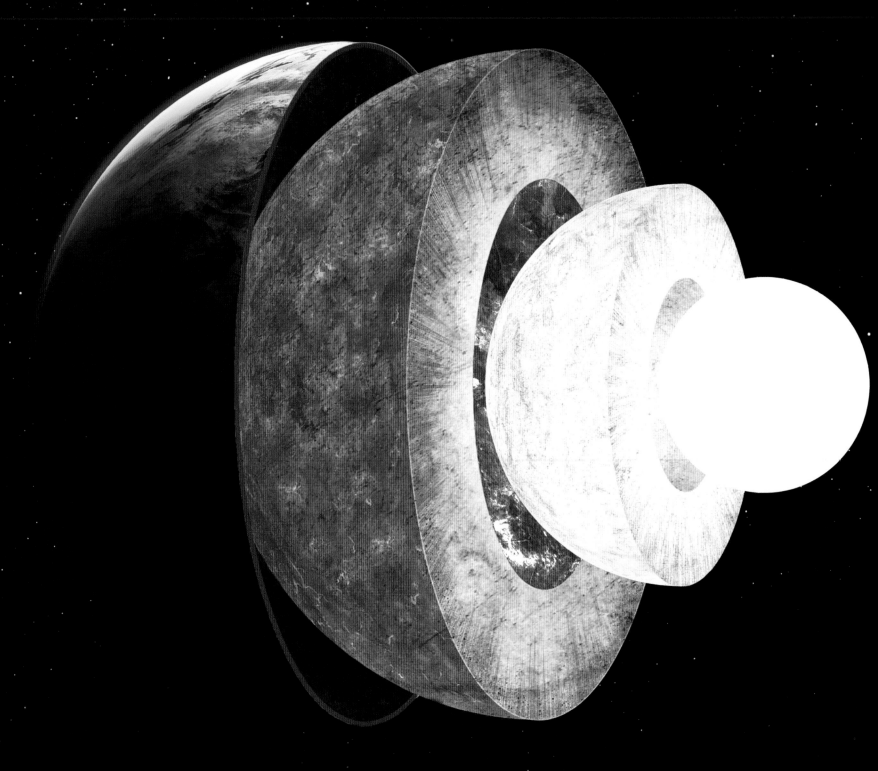

A simplified diagram of Earth's
structure comprising from right
to left: inner core, outer core,
mantle and crust.

Earth formed approximately 4.54 billion years ago with the most commonly-accepted hypothesis being one of accretion as matter from the solar nebula surrounding the infant Sun clumping together, with collisions generating heat to bind the matter into a growing rocky body able to exert an increasingly strong gravitational pull. Around 900 million years into Earth's life, it was struck by a large object approximately the size of Mars with the impact throwing out matter that later formed as Earth's sole satellite, the Moon.

The planet now consists of four distinct layers from the outer rocky crust to the thick mantle, a 1,400 mile-wide outer core containing fluid metals and the 1,518 mile diameter hot inner core made of solid iron and nickel and with temperatures approaching 9,800 degrees Fahrenheit. Earth's iron inner and outer cores generate a powerful magnetic field that extends tens of thousands of miles out into space.

Earth is the only Solar System planet with a rocky crust split into large chunks called tectonic plates. Seven large and a number of minor plates slowly shift, 'floating' on the plastic mantle that forms the asthenosphere beneath. Movement of the plates, slowly but with incalculable force over long time periods sees plates collide, subduct (one drives below another) and form mountain ranges such as the Andes and Himalayas. Volcanic and other geothermal activity is most marked on the boundaries between plates.

Earth's surface is divided roughly into a 70-30 split of water over land. Some 97 per cent of all the water on Earth is contained in its seas and five oceans which have an average depth of 2.5 miles and a deepest point in the Pacific of 36,070 feet at Challenger Deep. The remaining water, freshwater, is mostly locked into the polar ice gaps and glaciers leaving less than one per cent of all of the planet's water available to life in the form of rivers, streams, lakes, surface run-off and underground aquifers.

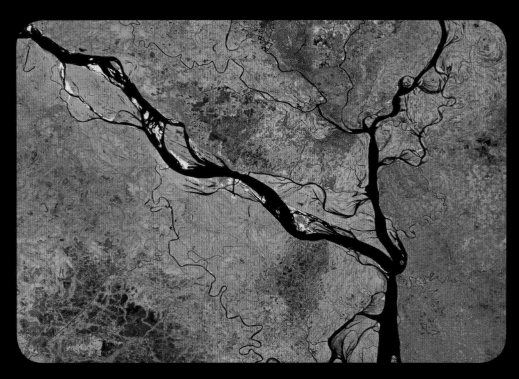

ABOVE: A satellite image from Landsat 8 captures the Padma River in India and Bangladesh and its numerous branches and tributaries. The Padma, along with many other rivers empty into the Bay of Bengal and shape the landscape along their route via erosion and deposition. Unique in the Solar System, Earth is the only known place where liquid water runs freely and extensively in the form of rivers and streams across the body's surface.

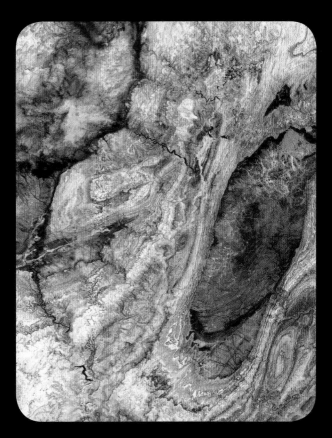

LEFT: This image, acquired by the Advanced Spaceborne Thermal Emission and Reflection Radiometer (ASTER) on board NASA's Terra satellite, depicts the Anti-Atlas Mountains of Morocco. These were formed as a result of the collision of the African and Eurasian tectonic plates about 80 million years ago and resulted in complex folding of rocks including underlying granite rocks (show in dark blue and green) and limestones and sandstones (yellow and orange).

Extending more than 60 miles above the land and sea's surface, the planet's thick, rich atmosphere comprises a cocktail of gases, most notably nitrogen (78.1 per cent), oxygen (20.9 per cent), argon, carbon dioxide and other, trace gases. Three quarters of the entire atmospheric mass is found within the lower 6.8 miles of the troposphere – the atmospheric layer closest to the planet's surface. Earth's atmosphere performs a series of invaluable tasks from shielding the planetary surface from potentially harmful radiation from space including gamma rays and high doses of ultra-violet radiation, to trapping and retaining heat, producing and distributing rainfall through winds as well as providing oxygen-rich air used by organisms for respiration.

The gas composition decreases in density the farther one travels upwards through the atmosphere, reaching the ultra-thin exosphere some 300 or more miles above Earth. At an altitude of 62 miles, the Kármán line is often used to delineate Earth's atmosphere from outer space whilst the effects of atmospheric re-entry are typically felt by manned and unmanned spacecraft descending to altitudes starting at around 75 miles.

Following exploration on land and by sea, large quantities of further knowledge about Earth has been gleaned when humans took to their air or sent unmanned drones, radiosondes and other scientific balloons, probes and satellites to investigate Earth from above. Peering down on the planet reveals just how diverse, extraordinary yet vulnerable our home world is. Earth possesses a remarkable range of different terrains, biomes and habitats for different living organisms – from cold taiga forests and shallow coral reefs to temperate grasslands and richly biodiverse, lush tropical rainforests.

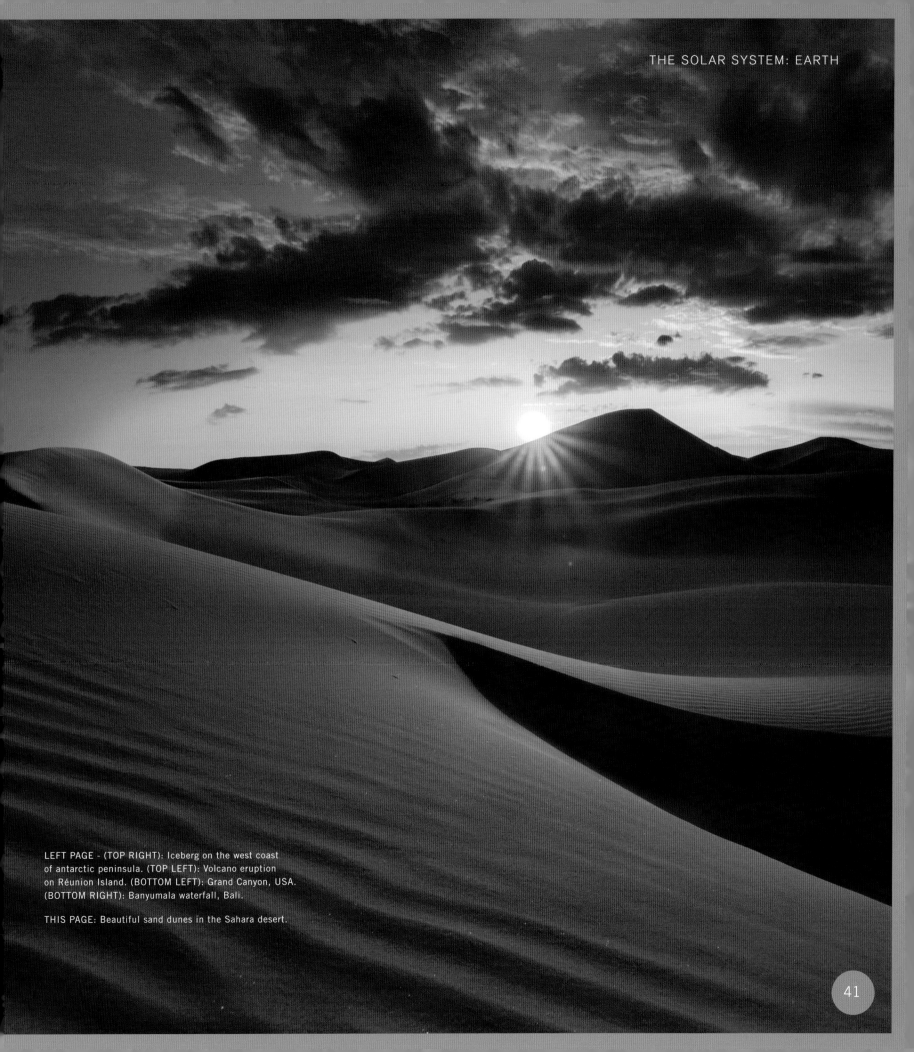

LEFT PAGE - (TOP RIGHT): Iceberg on the west coast
of antarctic peninsula. (TOP LEFT): Volcano eruption
on Réunion Island. (BOTTOM LEFT): Grand Canyon, USA.
(BOTTOM RIGHT): Banyumala waterfall, Bali.

THIS PAGE: Beautiful sand dunes in the Sahara desert.

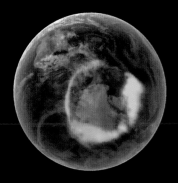

These ethereal, entrancing, and sometimes dramatic displays of shimmering and dancing brightly-colored lights occur high in the Earth's atmosphere and are best seen from relatively near the planet's poles.

AURORAE

THIS PAGE (TOP): From space, aurorae appear as crowns of light that circle each of Earth's poles. This view of the aurora australis (southern lights) was taken in 2005 by the IMAGE (Imager for Magnetopause-to-Aurora Global Exploration) satellite. The bright ring of light that a solar storm helped generate over Antarctica glows green in the ultraviolet part of the spectrum, depicted in this image.

ABOVE: An observation of Earth taken by International Space Station (ISS) Expedition 40/41 crew member Alexander Gerst reveals the aurora in stunning clarity.

To some native peoples in the past, they represented their ancestors flickering in the night sky. Science knows them as interactions between gaseous atoms in Earth's atmosphere and the hurtling solar wind from the Sun's surface.

A continuous flow of plasma made up of electrically charged particles emanates from the Sun's hot corona. This solar wind travels across space at a typical rate of 186 miles per second but its velocity and other properties can vary. Over coronal holes, it can reach velocities of 300-500 miles per second and reach temperatures of one million degrees Fahrenheit. Earth's magnetic field deflects most of the particles but, on occasion, some escape into the atmosphere where they collide with and excite atoms enough to emit photon energy in the form of light. Different atomic elements are responsible for the varying colors which tend to occur at different altitudes. Green lights, for example, appear at a typical altitude of 150 miles whilst purple and violet displays tend to occur from 60 miles upwards.

The Aurora Borealis or 'Northern Lights' can be viewed from many northerly locations in the northern hemisphere with popular congregating points including isolated regions of Alaska, northern Canada, Iceland and Scandinavia. The southern hemisphere equivalent, the Aurora Australis, offers equally stirring displays which are sometimes visible in southern Australia, New Zealand and the southernmost tips of Chile and Argentina. Aurorae, though, are not unique to Earth. They occur on other planets with the strong magnetic fields present in all four gas giant planets leading to prominent light displays despite their extreme distance from the Sun.

ABOVE: A striking Aurora
Borealis light display over
a fjord in Scandinavia.

43

Occasional but delightful light displays of individual 'shooting stars' can be viewed over a region of hundreds or thousands of square miles yet be caused by particles of matter smaller than a pea and often no bigger than grains of sand.

METEOR SHOWER

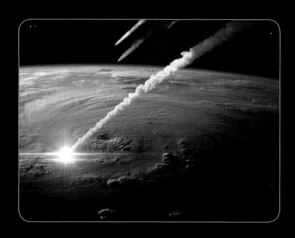

ABOVE: A meteor (bottom) from a meteor shower burns up whilst travelling through Earth's upper atmosphere in this image, parts of which were supplied by NASA.

These particular space visitors are known as meteoroids before they manage to enter the atmosphere, at which point they are dubbed meteors. Millions of meteors enter Earth's atmosphere each day. When a number occur and appear to emanate from the same or similar points in the night sky, they are collectively known as a meteor shower.

Many regular and established meteor showers occur each year as Earth's orbit crosses a debris path left behind, usually by a comet. These include the Geminids every December and the Draconids each October. One of the most vibrant and prolific displays is provided by the Perseids, so named as their radiant (the location from which they appear to come from) lies within the constellation Perseus. They are remnants of the dusty trail left behind by Comet 109/P Swift-Tuttle which last passed Earth in 1992 and won't return until 2126. Starting in mid-July and lasting to late August, meteor numbers typically peak on the 12th, 13th and 14th August each year with 100 or more meteors viewable in an hour which reach temperatures of 2,900 degrees Fahrenheit as they burn up 44 to 65 miles above Earth's surface. Particularly extreme or intense meteor showers, dubbed meteor storms, are rarer. On a roughly 33-year cycle, the Leonids meteor shower becomes a storm which can illuminate the November night sky with thousands of meteors per hour.

Geminid meteor shower in the night sky over Doi
Inthanon mountain, Chiang Mai, Thailand.

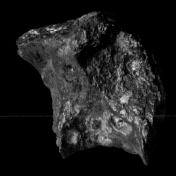

Rocks from space that survive an encounter with Earth's atmosphere and land on its surface are called meteorites.

METEORITES

THIS PAGE, TOP: Iron meteorite from the Barringer Meteor Crater in Arizona, USA. Weight 110g.

ABOVE: Hoba meteorite in Namibia - the world's largest intact meteorite yet discovered.

RIGHT: View of Meteor Crater (also known as Barringer Crater) near Flagstaff, Arizona. The crater has a maximum depth of some 560 feet whilst the crater rim rises 148 feet above the surrounding plains.

Whether chondrite (stony), made of metal (predominantly iron), or a mixture of rock and iron, many meteorites exhibit a burned exterior known as a fusion crust formed by the outer surface melting during atmospheric transit. Whilst more than 40,000 meteorites have been discovered on Earth, only one has been verified as directly striking and injuring a person – an Alabama resident, Ann Elizabeth Hodges in 1954. The eight and a half pound meteorite crashed through the roof of her Oak Grove home and caught her a glancing blow, causing bruising and making her a celebrity for a short period.

The largest surviving meteorite was discovered by Jacobus Hermanus Brits in 1920 whilst working the land of his farm near Grootfontein, Namibia. The Hoba West meteorite, dated at between 190 and 410 million years of age, and originally weighing around 70 tons is estimated to have landed on Earth around 80,000 years ago. America's largest meteorite, and the sixth largest in the world, was discovered in Oregon and named Willamette. Comprising 92 per cent iron and almost eight per cent nickel, this meteorite weighs 32,000 pounds.

The Hoba meteorite's large, flat sides may have acted as a significant brake on the rock on its descent, helping to keep it intact. Most large meteorites fragment during their atmospheric journey or break up on impact. Although destroyed themselves during impact, some meteorites leave behind giant impact craters, one of the best preserved of all being Barringer or Meteor Crater, near Flagstaff, Arizona. Formed by a 164 feet-wide meteor landing some 50,000 years ago, the crater measures almost 3,900 feet in diameter. One of the largest, verified craters is found in South Africa at Vredefort and measures 186.5 miles in diameter.

The outermost of the Solar System's four rocky planets, water once flowed across Mars' surface but it is now a cold, dry, dusty world.

MARS

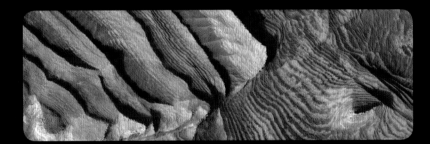

LEFT: Periodic layering of sedimentary rocks with trapped sand is exhibited in this image of the 104 mile diameter Becquerel Crater taken by Mars Reconnaissance Orbiter's HiRISE camera.

Possessing one tenth of the mass of Earth, Mars has a diameter of 4212 miles, around half the size of Earth, and contains a heavy core made of iron, nickel and sulfur surrounded by a rocky mantle and topped by a solid crust. Oxidation of iron compounds in the Martian soil is responsible for its distinctive red color that caused the ancient Chinese to label it the 'Fire Planet' and the Ancient Egyptians to name it Her Desher meaning the Red One.

Taking 687 Earth days to complete its orbit at an average distance of 142 million miles from the Sun, Mars possesses a thin atmosphere comprising over 95 per cent carbon dioxide with small amounts of nitrogen and argon. This barely mitigates the planet's large temperature swings from as low as -284 degrees to as high as 86 degrees Fahrenheit according to NASA's Mars Exploration Program. The planet's two small and irregular potato-shaped moons measure 13.8 miles (Phobos) and 7.8 miles (Deimos) across. Discovered by Asaph Hall in 1877 and named after twins in Greek mythology, these may have been originally asteroids captured by Mars' gravity which is 38 per cent of that experienced on Earth. With its orbit gradually decreasing in altitude, Phobos is considered doomed to crash into Mars within 30 to 50 million years.

Amongst Mars' most superlative high and low points across its 55.7 million square miles of surface area is the Solar System's biggest mountain. Olympus Mons. Measuring over 300 miles across, this giant shield volcano rises 85,000 feet above the surrounding plains and some 72,000 feet above the planet's mean surface level as measured by the Mars Global Surveyor's MOLA (Mars Orbiter Laser Altimeter) instrument. In contrast and with a depth of 23,465 feet, Hellas Planitia is one of the Solar System's largest and deepest visible impact craters.

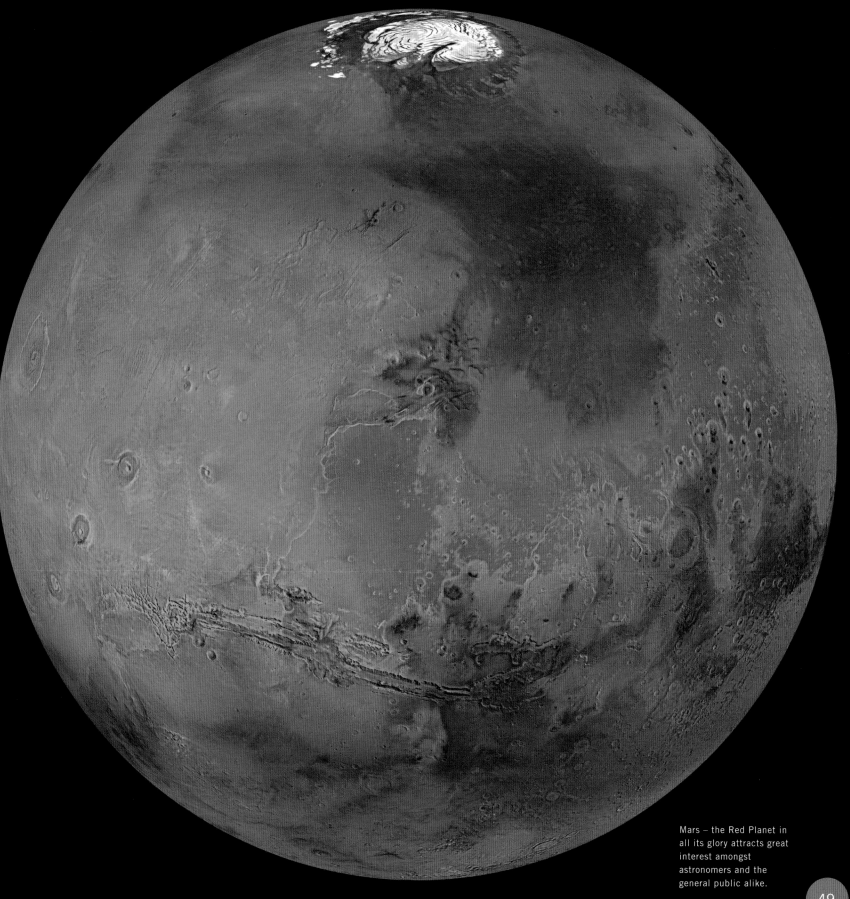

Mars – the Red Planet in all its glory attracts great interest amongst astronomers and the general public alike.

Whilst canyons and volcanoes excite much public interest, more of Mars' surface is formed of large flat or gently sloping plains than any other terrain. Rugged highland plains are called plana whilst lowand plains are known as planitiae and are extensive.

MARTIAN PLAINS

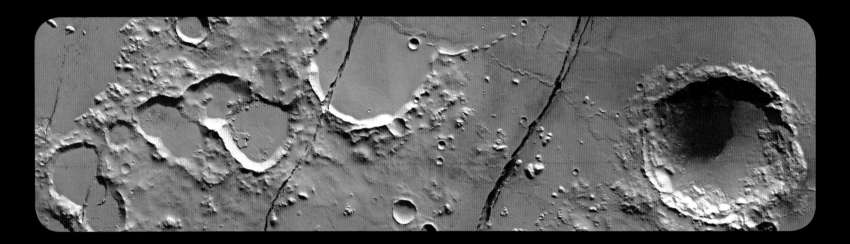

The northern lowland plains of Mars, for example, cover almost a third of the planet and make up its largest lowland regions. Together, highland and lowland plains cover more than 60 per cent of the planet's surface.

Mars' plains have formed in a variety of ways. Chryse Planitia, for example, is thought to have once been a giant impact basin, formed two to three billion years ago, which has since been eroded to form a smooth circular plain, approximately 1,000 miles in diameter and at an elevation one and a half miles

below the planetary average. It was first investigated by the pioneering Viking I lander between 1976 and 1982 and in 1997 by the first moving robot on another planet – NASA's Pathfinder-Sojourner mission which found evidence of water erosion in the distant past. Argye Planitia is a similar impact basin-formed plain, some 3.9 billion years old and depressed more than three miles below plains areas that surround it.

Other plains, such as Utopia Planitia and Amazonis Planitia were the result of volcanic action with molten lava spewing

out from vents and flowing long distances across the surface, particularly in the northern hemisphere of the planet. Many of these volcanic plains formed more than two billion years ago, but Amazonis Planitia is young, at only 100 million years old and one of the smoothest and most level plains on the planet. The rock-strewn flat terrain of Utopia Planitia was chosen as the landing site for the Viking II mission. It detected basalt rocks riddled with holes where bubbles of volcanic gas had burst within the rock and an ultra-thin layer of frost, less than a millimetre thick.

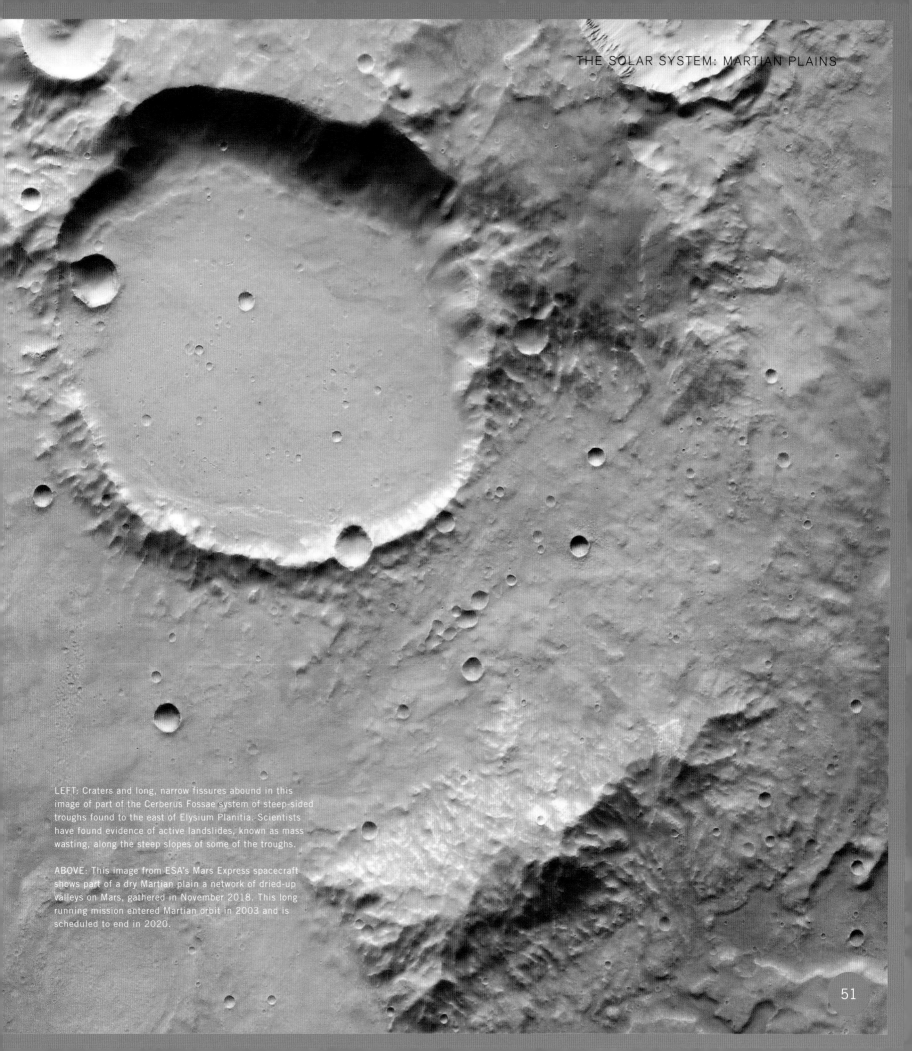

LEFT: Craters and long, narrow fissures abound in this image of part of the Cerberus Fossae system of steep-sided troughs found to the east of Elysium Planitia. Scientists have found evidence of active landslides, known as mass wasting, along the steep slopes of some of the troughs.

ABOVE: This image from ESA's Mars Express spacecraft shows part of a dry Martian plain a network of dried-up valleys on Mars, gathered in November 2018. This long running mission entered Martian orbit in 2003 and is scheduled to end in 2020.

Located just north of the equator, Elysium Planitia measures approximately 1,700km by 2,400km and contains mesas and buttes as well as river valleys including Athabasca Valles – estimated at between just four and million years old, making it one of the youngest valley formations on the Red Planet. In 2005, the Mars Express spacecraft revealed the possibility that a substantial part of the plain (an estimated 560 miles by 500 miles or approximately the size of the North Sea on Earth) may house a frozen sea covered by dust which protects the ice from vaporizing into the atmosphere. Estimated to reach 160 feet in depth, the ice may be the remains of floods from fissures in the Martian surface.

Mars' rugged highland plains and plateaus are mostly found in the southern hemisphere. Some plana show signs of stresses in Mars' crust which yield geological features such as pit chains and grabens – blocks of land bordered by faults in the Martian crust which has seen them slip lower than the surrounding surface. Grabens exist in Meridiani Planum which also includes layered sedimentary rocks containing hematite, a form of iron ore. On Earth this is mostly found in locations where water has been present, leading to speculation that the plain may have been the site of a large lake or sea over 3 billion years ago.

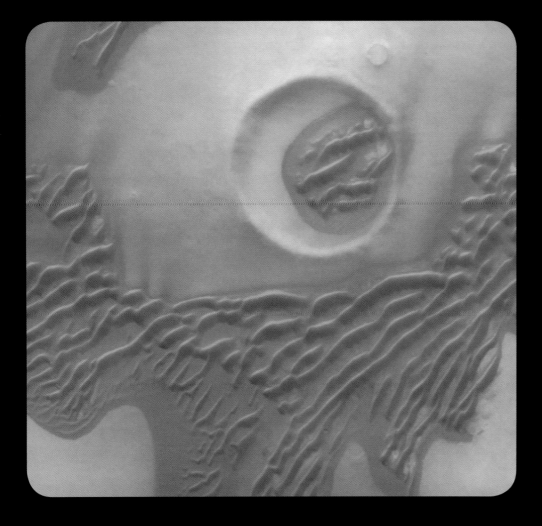

ABOVE LEFT: The desolate rocky terrain of Utopia Planitia, found in Mars' northern hemisphere and the landing site of the Viking 2 probe in 1976.

LEFT: Not all Martian sand dunes are located in craters. This image from the Mars Odyssey spacecraft in 2001 shows dunes located on the plains of Terra Sirenum, located in the planet's southern hemisphere.

Pathfinder on Mars.

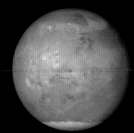

The Solar System's biggest canyon utterly dwarfs America's Grand Canyon in scale and complexity. Valles Marineris is around nine times longer and five times deeper than Arizona's pride and joy.

VALLES MARINERIS

Found close to the planet's equator, near the Tharsis region of Mars, the meandering canyon sprawls over a length of 2,500 miles – approximately the straight line distance between New York City and Los Angeles – meaning it extends almost a quarter of the way around the planet. Parts of this canyon system descend beyond 23,000 feet whilst in places the canyon extends to 120 miles in width, forming a huge geological scar across the surface of the planet. Some of the canyon walls themselves are cut with deep gullies, indicating that erosion has

been at work whilst closer examination of the canyon floors reveal heavy depositions of matter in places leading some to speculate that water once flowed through the giant chasms.

The canyon system stretches from the Noctis Labyrinthus--a system of maze-like valleys and canyons in the west--to the chaotic the chaotic terrain near the Chryse Planitia basin. Huge ancient river channels began from Valles Marineris and from adjacent canyons and ran north. Many of the channels

flowed north into the Chryse Basin. Valles Marineris was named after the Mariner 9 spacecraft which discovered the canyon system on its mission to Mars in 1971 when it became the first space probe to go into orbit around another planet. Different regions and canyons within the system have been named, including Coprates Chasma to the east--a 600-mile-long canyon containing multiple recently-discovered volcanoes, some with cones stretching 1300 feet high.

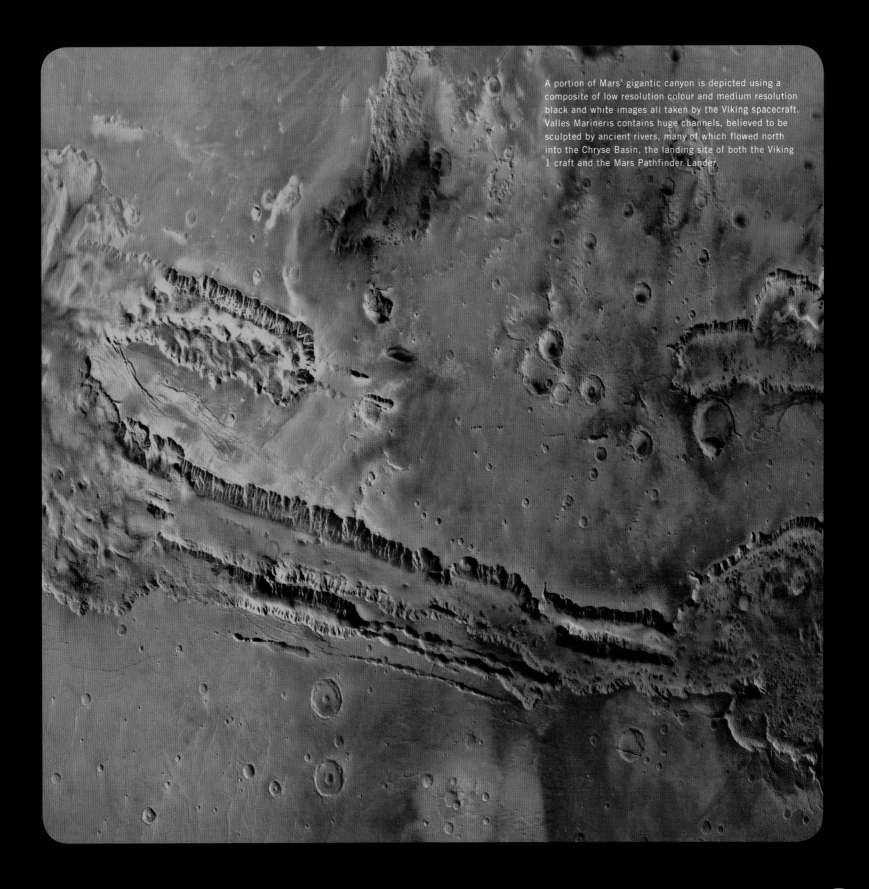

A portion of Mars' gigantic canyon is depicted using a composite of low resolution colour and medium resolution black and white images all taken by the Viking spacecraft. Valles Marineris contains huge channels, believed to be sculpted by ancient rivers, many of which flowed north into the Chryse Basin, the landing site of both the Viking 1 craft and the Mars Pathfinder Lander.

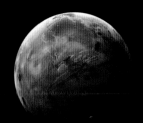

At its poles, the Red Planet
exhibits a familiar similarity
to our own planet.

MARS POLAR ICE CAPS

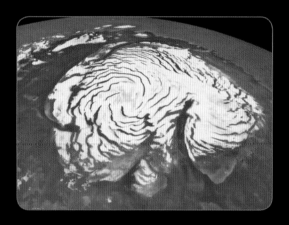

LEFT: Combining data from two instruments
on board NASA's Mars Global Surveyor, this
image depicts an orbital view of the north
polar region of Mars. To the right of center,
a large canyon, Chasma Boreale, almost
bisects the white ice cap.

A large accumulation of white hails the presence of ice in vast quantities which has excited scientists ever since telescopes enabled them to study the regions. The planet has an average mean temperature of -85 degrees Fahrenheit according to the NSSDC planetary factsheet, with temperatures lower at its poles, meaning that both poles are covered in permanent ice caps. This mostly consists of water ice some dust mixed with water in places to give patterned layers of bright 'clean' and darker 'dirty' ice. Just like Earth, there is variation in temperature during the seasons and as the Martian winter approaches and temperatures drop, so carbon dioxide from Mars' thin, tenuous atmosphere freezes, forming a layer of solid CO_2 building up the ice caps. As winter transitions into summer, the carbon dioxide will sublimate, changing state from a solid straight to a gas; between three to four trillion tons of carbon dioxide – as much as a seventh of Mars' entire atmosphere - undergoes these processes each year.

The northern polar ice cap is the larger of the two, measuring around 600 miles in diameter and rising at its thickest point some 10,000 feet above the surrounding plains. The southern ice cap has a similar thickness but smaller extent, with a diameter of some 250 miles. In 2018, a subglacial lake 12 miles across was discovered under part of the southern ice cap by a penetrating radar instrument, MARSIS, aboard the European Space Agency's Mars Express orbiter.

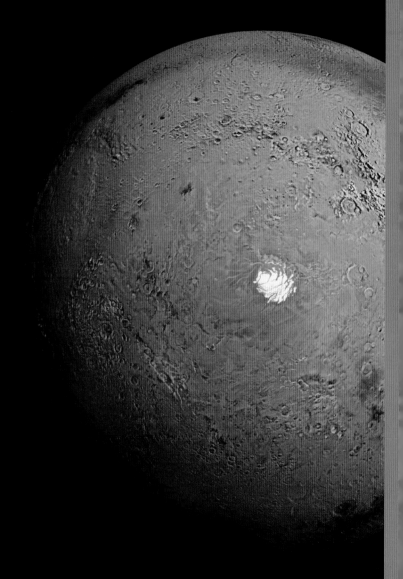

ABOVE: Artist impressions of Mars northern (left) and southern (right) polar ice caps. Surrounding the southern ice cap are a large field of eskers – ridges made of sediment and debris deposited by meltwater from a retreating ice sheet or glacier and potential evidence that the southern polar ice cap once covered a far greater extent.

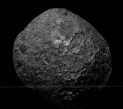

Rocky remnants of the Solar System's formation, these small bodies mostly occupy a thick belt lying between the orbits of Mars and Jupiter.

ASTEROIDS

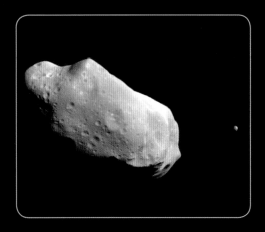

TOP: The SUV-sized Asteroid 2008TC3 was tracked into Earth's atmosphere where it broke up some 23 miles above the Nubian Desert in Sudan. Some 600 small meteorites were later recovered on the ground.

ABOVE: During its examination of the asteroid Ida, NASA's Galileo spacecraft returned images of a second object, Dactyl - the first confirmed satellite or moon of an asteroid. Dactyl is just about visible to the right of Ida.

The largest, named Vesta after the Roman goddess of the home and hearth, measures 329 miles across, but the vast majority are far smaller. NASA estimated that the belt contains between 1.1 and 1.9 million asteroids larger than 0.6 miles. Yet, of this large number, less than 25 have a diameter of 125 miles or more. Combining all the belt's asteroids would only yield a mass around one-tenth that of the Moon. Despite being relatively small, more than 150 asteroids in the belt are known to be accompanied by their own small moon or moons.

Asteroids are found elsewhere. The Trojans, for example, are clusters of asteroids which travel ahead or behind of Jupiter's orbit around the Sun. Other asteroids are found deep within the Kuiper Belt. In contrast, some asteroids lurk closer or cross our planet's orbital path. Near Earth Asteroids (NEAs) are defined as those within 28 million miles of Earth. The JPL Center for NEO Studies (CNEOS) and other agencies track NEAs. There were over 19,300 by 2019, but only a handful are considered to pose any real potential threat.

In 2014, a 1200-foot-wide asteroid, 2014 HQ124, travelled past our planet within three and a quarter times the distance between Earth and the Moon – this was considered a near miss.

Many missions to asteroids have been performed or are scheduled such as NASA's Psyche mission, slated for 2022, to explore an all-metal asteroid and the year before, the Lucy mission which will embark on a 12 year journey to Jupiter to become the first probe to investigate the Trojans. One of the most audacious was Japan's Hayabusa spacecraft which piggy-backed onto the main belt asteroid, Itokawa in 2005 for 30 minutes and sent back a sample return capsule which reached Earth five years later.

BELOW: This panoramic view of the giant asteroid Vesta was taken by NASA's Dawn spacecraft, as part of a rotation characterization sequence in July 2011, at a distance of 3,200 miles. It shows numerous impact craters of various sizes and grooves parallel to the equator.

LEFT: This montage shows the asteroid 951 Gaspra (top), discovered in 1916, with dimensions of approximately 12.5 x 7.5 x 7 miles as compared with the two moons of Mars - Deimos (lower left) and Phobos (lower right).

Discovered by an Italian Catholic priest, Father Giuseppe Piazzi, in 1801, Ceres is the largest body in the Asteroid Belt, the broad ring lying between Mars and Jupiter containing at least 800,000 asteroids over 0.6 miles in diameter.

CERES

With a diameter of 585 miles, Ceres comprises approximately 25-35% of the mass of the entire belt and has been classified as a dwarf planet – the only such example in the inner Solar System. Located an average of 257 million miles from the Sun, Ceres takes 4.6 years to complete its orbit. Many astronomers consider it a protoplanet that didn't quite make it, possessing an oblate spherical shape, unlike its irregular asteroid neighbours in the belt, but lacking the gravitational force to clear its orbital path free of other bodies.

In 2015, the Dawn space probe powered by an innovative ion propulsion system approached Ceres and went into orbit around the dwarf planet the following year, getting as close as 22 miles above its surface. It revealed a heavily cratered but relatively uniform body with one major pyramidal mountain, Ahuna Mons, reaching approximately 16,000 feet in elevation. In some craters, notably the 50 mile-wide Occator crater, bright white patches called faculae reflect sunlight. These consist of deposits of carbonates and other salts suggesting a watery past; Ceres' low density also suggests the presence of much ice in shadowed craters and below the dwarf planet's outer surface.

LEFT: This view of Ceres, produced by the German Aerospace Center in Berlin, combines images taken with different spectral filters by the Dawn spacecraft to give an impression of the dwarf planet's coloration to use if viewed directly.

BELOW: A stretch of Ceres' southern hemisphere, taken by the Dawn spacecraft shows its gently undulating terrain pock-marked with young and mostly small craters. Observations have revealed the presence of graphite, sulfur, and sulfur dioxide on Ceres's surface.

Named after the Roman king of the gods and the third brightest object in the night sky after the Moon and Venus, Jupiter was known as Marduk in ancient Babylonia and Zeus in ancient Greece.

JUPITER

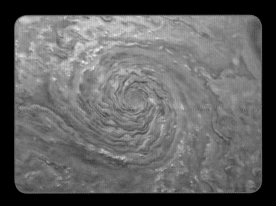

ABOVE: A cyclonic storm in Jupiter's northern hemisphere is captured in this image from NASA's Juno spacecraft. Many bright white cloud tops can be seen popping up in and around the arms of the rotating storm. The color-enhanced image was taken in February, 2019, as the spacecraft performed its 17th science flyby of the planet.

With a diameter of 86,881.4 miles or 11 Earth widths, it is the largest planet in the Solar System and more than twice as massive as all the other planets put together with a correspondingly large volume equivalent to 1,300 Earths. A common visual guide to scale is to compare a basketball as Jupiter with a nickel as Earth. For such a large body, Jupiter is a fast mover, rotating on its axis at a speed of 27,961 mph and completing a full rotation in just 9.94 hours. The planet's rapid rotation is behind a 5,716 mile bulge around its middle. Four broad but faint rings encircle the planet, predominantly consisting of dust.

This gas giant has no solid surface but one made of cold swirling gases and liquids at a surface temperature of -166 degrees Fahrenheit. From our viewpoint, the top of Jupiter's atmosphere features striking bands known as belts (sinking regions of gas) and zones (rising regions) which extends from the surface down approximately 44 miles. Fast winds sweep and churn the atmosphere at speeds rising typically to 225mph whilst Jupiter's giant magnetosphere (which extends more than two million miles in the direction of the Sun and tapers but lengthens to reach Saturn's orbit in the opposing direction) interacts with charged particles from space to form spectacular aurora at or near the planet's poles with 100 times more intensity than aurora found on Earth.

ABOVE: Jupiter's complex
tapestry of cloud bands,
atmospheric layers and giant,
rotating atmospheric storms
can be clearly viewed in this
3D rendering using image
elements from NASA.

Jupiter's most iconic feature, this swirling storm is on an unparalleled scale.

GREAT RED SPOT

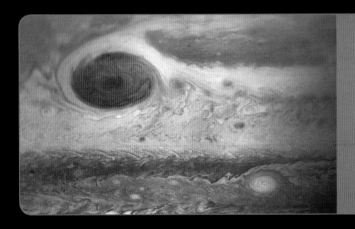

A color-enhanced image of the iconic Great Red Spot taken by the Juno spacecraft as it travelled between 15,379 and 30,633 miles above the tops of the planet's clouds.

Juno entered a polar orbit around Jupiter in 2016 with an expected end of mission date in July 2021. Its measurements and data gathering concerning the Great Red Spot and other storms is hoped to help scientists understand further the planet's complex weather systems.

NASA's Juno spacecraft made accurate measurements of the spot's current dimensions in 2017, denoting its width as 10,159 miles, making it approximately 1.3 times as wide as Earth. The storm's cloud tops extend some five miles higher than the surrounding Jovian atmosphere but descend an incredible 200 miles or so beneath the atmospheric surface. Yet, these super-sized figures, impressive as they may sound, indicate that the storm has shrunk significantly since it was first observed and measured. It once spanned more than two Earth-widths in size whilst the visiting Voyager 2 spacecraft in 1979 measured the storm at 14,500 miles wide. Like many aspects including the storm's precise origins, why its coloration is so marked and the detailed chemical make-up of its swirling clouds, the storm's shrinking remains an alluring mystery.

The Great Red Spot (GRS) is a violent anticyclone located in Jupiter's southern hemisphere which takes about six Earth-days to completely rotate in a counter-clockwise direction. A relatively calm centre belies its outer layers where winds rage at a maximum speed of between 270 and 425 miles per hour. Besides its great extent and violence, the GRS's most extraordinary attribute is its great longevity. The Great Red Spot has been continuously studied in a structured fashion since 1878 when it was described by Missouri-based astronomer Carr Walter Pritchett, but it is thought to have raged for centuries before and may have been observed as far back as the 17th Century by astronomers such as the Italian, Giovanni Cassini.

Another arresting view of Jupiter's Great Red
Spot (top right) along with the turbulent
atmosphere of the planet's southern
hemisphere, taken by the Juno spacecraft,
in February 2019. Clearly visible are several
of the 'string of pearls' – counterclockwise
rotating giant storms appearing as white ovals.

When Italian scientist Galileo Galilei trained his homemade telescope onto the vicinity of Jupiter in 1610, little did he know that he would be kick-starting a long process.

MOONS
OF JUPITER

TOP, FROM LEFT TO RIGHT: The four Galilean moons of Jupiter: Callisto, Io, Ganymede and Europa. These four dwarf all other Jovian moons in scale, possessing diameters at least 12 times greater than the next largest moons.

ABOVE: Jupiter looms over its moon, dwarfing Ganymede – the largest satellite in the Solar System – in this enhanced-contrast image from the Cassini spacecraft.

Galileo discovered four moons which German astronomer Simon Marius named Io, Europa, Callisto, and Ganymede. In 2018, a team led by Scott S. Sheppard of Carnegie University discovered 12 new moons of Jupiter, taking the known total to 79, and making Jupiter the most populous planetary satellite system. Ganymede is the largest moon in the Solar System with a diameter of 3,273 miles, around two-thirds the size of Mars. Its landscape is a mix of dark older plains and young craters through which has seeped slushy water ice. Europa is the smallest of the four Galilean moons and with a crust largely consisting of water ice criss-crossed by countless pinkish-brown lines which many scientists believed are formed of ice mixed with hydrated salts such as magnesium sulfate and sulfuric acid. The crust, between 10 and 15 miles deep, is posited to float on a subterranean ocean, an exciting possibility to be investigated by the potential Europa Clipper mission slated for a 2023 launch. Io is Jupiter's innermost major moon. It travels around Jupiter in a 42.5 hour, slightly elliptical, orbit and is engaged in a gravitational tug-of-war with its parent planet and the other moons. The result is a stretching and squeezing of its crust leading to prominent and extensive volcanic activity across its surface. In marked contrast to Io, Callisto is the most inactive of the four Galilean moons. Its ancient surface, as old as four billion years, is saturated with meteorite scars and innumerable craters.

TOP, FROM LEFT TO RIGHT: Callisto and Io. ABOVE, FROM
LEFT TO RIGHT: Ganymede and Europa. (Elements of this
image/illustration furnished by NASA).

67

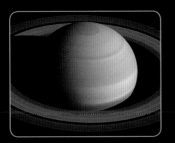

Named after the ancient Roman god of agriculture, the sixth planet from the Sun is renowned for its rings and shares a lot of similarities with its bigger brother, Jupiter.

SATURN

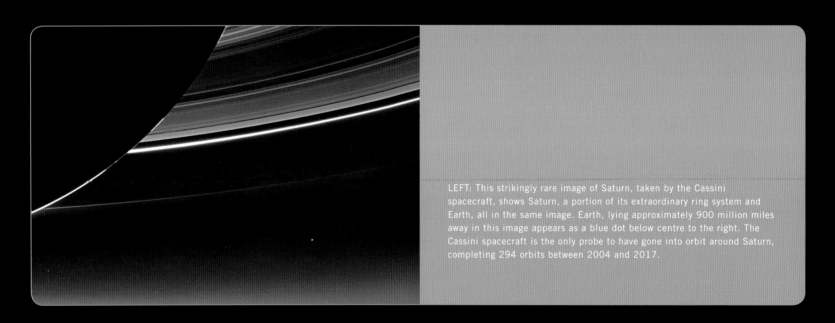

LEFT: This strikingly rare image of Saturn, taken by the Cassini spacecraft, shows Saturn, a portion of its extraordinary ring system and Earth, all in the same image. Earth, lying approximately 900 million miles away in this image appears as a blue dot below centre to the right. The Cassini spacecraft is the only probe to have gone into orbit around Saturn, completing 294 orbits between 2004 and 2017.

Both are gas giants that dwarf all other Solar System bodies (save for the Sun) and both spin rapidly leading to a bulge around their equator, 7,337 miles in Saturn's case. Saturn lies an average of 886 million miles away from Earth and takes 10,747 days to complete its orbit according to the NSSDC. The planet is around nine times as wide as Earth and is thought to feature a rocky and metallic core, small compared to the planet's size but still, according to the

European Space Agency, around 10 times the mass of Earth. The remainder of the planet is predominantly helium and especially hydrogen which is gaseous in its outer layers but beneath the surface, huge pressures transform the gas into a metallic liquid layer. Saturn has the lowest density of any planet in the Solar System. At 0.248 pounds per cubic inch, it is eight times less dense than Earth and would float in water.

The outer atmosphere of Saturn contains 96.3% molecular hydrogen and 3.25% helium by volume with trace amounts of ammonia, propane, ethane, methane and acetylene present. This is whipped into bands by powerful winds which reach speeds of up to 1,600 feet per second in the equatorial region. Occasionally, violent 'white' storms break through the cloud layers, each one bigger than Earth in diameter.

ABOVE: Two whole planet views of Saturn and its
magnificent ring system depict the planet from above
and below its equatorial plane.

Beyond its atmosphere, Saturn is orbited by 62 moons, some of which, such as Prometheus and Pandora, are irregularly-shaped.

SATURNIAN MOONS

Astronomers are particularly absorbed by Titan, the largest moon of Saturn and its liquid ocean as well as the unusual and partially hollow Hyperion – the first non-spherical moon to be discovered (in 1848). Its outer surface looks more akin to a sponge than a rocky moon with legions of sharp vertical craters. Arguably, the most fascinating Saturnian moon of all is Enceladus with its cryovolanoes that spew jets of water vapour, ice, hydrogen and salt crystals out into space at speeds of 800 mph.

Mimas is the innermost moon, a mere 246 miles in diameter, and orbiting Saturn at an average distance of 115,289 miles. The smallest spherical moon in the Solar System, its most prominent feature is a deep circular impact crater about a third of the width of the moon, giving it an distinctive appearance, reminiscent of the Death Star space station in Star Wars. The orbits of two small moons, Janus and Epimetheus (about 110 miles and 75 miles wide respectively) are a mere 30 miles apart but in a complex cosmic dance, the two never collide but exchange angular momentum every four years and switch orbits, an extraordinary phenomenon played out over 100 days and witnessed by the Cassini spacecraft in 2005-06.

Three moons, Tethys, Calypso and Telesto, the latter two small moons both discovered in 1980, are 'co-orbital satellites' sharing the same orbital path with Telesto some 60 degrees ahead of the far bigger Tethys and Calypso some way behind. Tethys, itself, is predominantly composed of water ice and features the Odysseus Crater – a 250 mile wide crater about two fifths the width of the moon as well as a giant 62 mile wide valley, Ithaca Chasma, that runs 1,200 miles from north to south across the moon's surface.

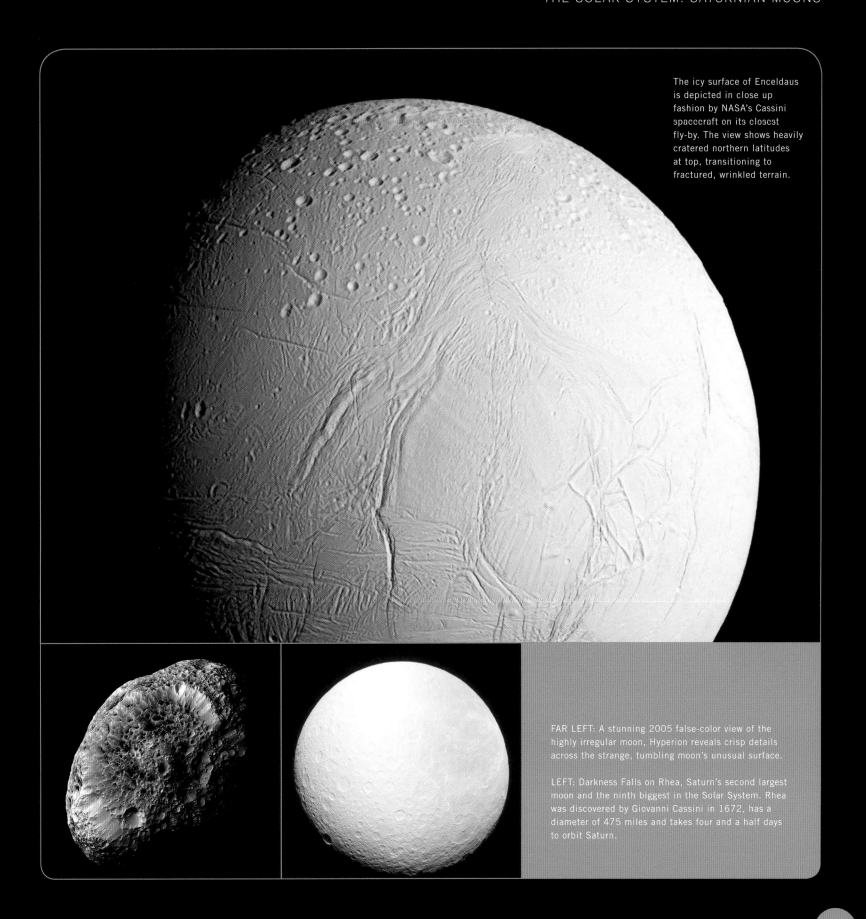

The icy surface of Enceldaus is depicted in close up fashion by NASA's Cassini spacecraft on its closest fly-by. The view shows heavily cratered northern latitudes at top, transitioning to fractured, wrinkled terrain.

FAR LEFT: A stunning 2005 false-color view of the highly irregular moon, Hyperion reveals crisp details across the strange, tumbling moon's unusual surface.

LEFT: Darkness Falls on Rhea, Saturn's second largest moon and the ninth biggest in the Solar System. Rhea was discovered by Giovanni Cassini in 1672, has a diameter of 475 miles and takes four and a half days to orbit Saturn.

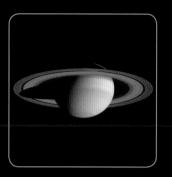

All four outer planets are encircled by systems of rings but Saturn's are by far the most prominent, beautiful and entrancing.

SATURN'S RINGS

In this striking image from the Cassini orbiting spacecraft, three of Saturn's small moons, Epimetheus and the two F ring shepherd moons, Prometheus and Pandora, can be clearly seen. Their proximity to the material forming the rings is thought to have the effect of sculpting the rings: giving them sharp edges, and helping to create gaps between them.

As Dava Sobel writes in her book, Planets, "Seen for the first time through a backyard telescope, ringed Saturn is the vision most likely to turn an unsuspecting viewer into an astronomer forever." The seven main rings, labelled A to G by astronomers, extend thousands of miles away from the planet.

Despite appearing uniform in telescope imaging, the rings are not solid discs but comprise billions of piece of debris ranging in size from smaller than a pea fragments to boulders the size of a house and very occasional giant pieces of ice more than a mile high. These particles are predominantly water ice with some rocky traces all bound together in a flat plane by Saturn's gravity. Shepherd moons including Pan, Atlas and Pandora lie within or close to the rings. Their gravity helps sweep their orbital paths of debris herding the icy particles into formation.

Saturn is tilted at an angle of 26.7 degrees meaning that it displays different aspects and angles of the rings to Earth throughout its orbit. The rings are seen edge-on from Earth every 15 years, and hence seem to disappear. Gaps between the rings are named after famous astronomers of the past including the Encke and Cassini Divisions, the latter measuring 2,920 miles wide. In 2004, the Cassini space probe slipped through the gap between Saturn's F and G rings to become the first spacecraft to orbit the planet.

This visible light image taken by NASA's Cassini spacecraft and released in September 2016, showcases aspects of the incredibly detailed structure of Saturn's rings and how they are formed from numerous ringlets which blur together when viewed from a distance. This view targets the sunlit side of the rings from a distance of 283,000 miles and about four degrees above the ring plane.

Saturn's largest moon is bigger than the planet Mercury and fascinates astronomers due to its range of unusual features.

TITAN

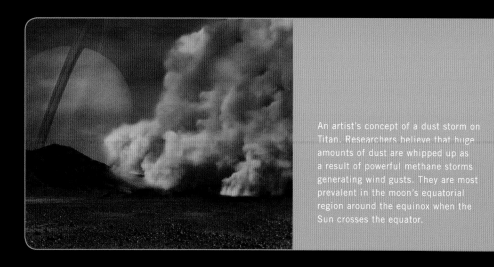

An artist's concept of a dust storm on Titan. Researchers believe that huge amounts of dust are whipped up as a result of powerful methane storms generating wind gusts. They are most prevalent in the moon's equatorial region around the equinox when the Sun crosses the equator.

TOP: Three views of Titan compiled from data gathered by the Cassini spacecraft on its last three fly-bys of the moon.

It is the only known moon in the Solar System to possess a substantial atmosphere and the only body, apart from Earth, to contain a nitrogen-rich blanket of gases surrounding its rocky surface. Discovered in 1655 by Dutch scientist, Christiaan Huygens, Titan's atmosphere is approximately 97 per cent nitrogen but with one and a half times the surface pressure of Earth's atmosphere. Its dense nature and distance from the Sun means that little sunlight penetrates and reaches the surface which experiences an average temperature of -290 degrees Fahrenheit. This icy body completes an orbit of Saturn once every 15 days, 22 hours.

Pioneer 11 and both Voyager spacecrafts performed fly-bys of the 3199.7 mile-diameter moon in a flurry of activity between 1979 and 1981, but the Cassini-Huygens mission was tasked with getting closer to Titan than ever before. In 2004, the Cassini orbiter reached Saturn and the following year released a small lander probe, Huygens, which descended through Titan's atmosphere for 147 minutes before landing on the moon's surface and transmitting observations and data for a further 90 minutes. Amongst Cassini and Huygens' discoveries were large lakes or seas, known as mares, containing flowing hydrocarbons including ethane and methane. The largest, Kraken mare, covers an area of approximately 150,000 square miles, making it the size of the Caspian Sea. The second largest, Ligeia Mare, is thought to be more than twice the size of Lake Michigan.

An infrared composite image of
Titan from the Cassini spacecraft's
VIMS (visual and infrared mapping
spectrometer) cuts through the
moon's hazy atmosphere to reveal
the dune-filled Fensal region
to the north. Just above center
lies Menrva, the moon's largest
confirmed impact crater.

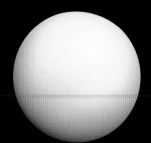

William Herschel initially thought he had spied a comet in 1781. What actually caught his eye was the first planet discovered using a telescope.

URANUS

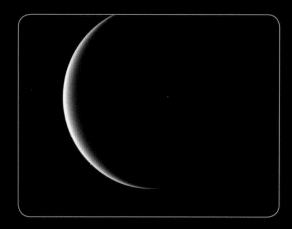

ABOVE: Uranus as seen by NASA's Voyager 2 spacecraft as it made its closest pass by Uranus, some 50,600 miles from the planet's cloudtops.

LEFT: Mostly in shadow, only a sliver of Uranus is seen in this image of Uranus taken by Voyager 2. The planet orbits the Sun at an average distance of 1,783,744,300 miles and is approximately four times larger than Earth.

Uranus is a cold gas giant an average of 19.8 AU away from the Sun and receiving just 1/400th of the Sun's energy that strikes Earth. Methane crystals in its upper atmosphere absorb red light giving the planet its distinctive green-blue color but its atmospheric gases are mostly helium and hydrogen with hydrogen sulfide confirmed present by astronomers in 2017. The 15,759.2-mile radius planet completes its orbit in 30,687 days, a shade over 84 years according to NASA, but spins about its axis rapidly, completing a rotation in 17 hours. Whilst all the Solar System planets exhibit some degree of axial tilt, Uranus is far and away the most extreme. An ancient collision or series of collisions billions of years ago may have been responsible for the planet effectively 'lying' on its side with an axial tilt of 97.7 degrees. As a result, the planet's poles are either plunged into darkness or receive continuous sunlight for long periods, even years at a time.

In 1977, whilst flying on board the Kuiper Airborne Observatory, three Cornell University astronomers, James Elliot, Ted Dunham, and Jessica Mink, discovered Uranus' ring system. Further investigations have revealed that the planet is surrounded by at least 13 narrow and mostly dark rings which encircle the planet's tilted equator and is orbited by 27 moons, the majority under 70 miles in diameter and all named after characters in literature by William Shakespeare or Alexander Pope. Discovered by Herschel, Titania it the largest moon at 981 miles in diameter, whilst Miranda, discovered by Gerard Kuiper in 1948, contains the Solar System's tallest cliffs named Verona Rupes and towering 12 miles high.

Uranus viewed whole in this artist's impression using imagery from NASA. The planet is the coldest in the Solar System with temperatures recorded as low as -371 degrees Fahrenheit.

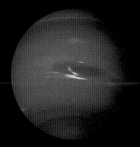

Named after the Roman god of the sea for its ethereal mid-blue coloration, Neptune is the most distant Solar System planet from the Sun, orbiting an average of 2.79 billion miles away - more than 30 times the distance Earth is from the Sun.

NEPTUNE

LEFT: A dramatic view of a crescent-like portion of Neptune acquired by the Voyager 2 spacecraft in the summer of 1989 shortly after it had come to within just 3000 miles of the planet's north pole.

At such an extent, the planet receives 900 times less of the Sun's energy than Earth and as a result its surface temperatures average -330 °F across the 30,775 mile diameter gas giant. Most of the planet's mass in its interior is actually a dense fluid of 'icy' materials including water, methane and ammonia, all of which surround a small, rocky core.

Neptune rotates rapidly, completing its day in 16 hours but its year is 165 times longer than Earth's. Our knowledge of

the planet was relatively scant until it was transformed by the 1989 fly-by of NASA's Voyager 2 deep space probe. The probe and subsequent investigations reveal the most windswept world in the Solar System with gusts of more than 1,200 mph in its atmosphere and vast storms, one of which, the Great Dark Spot, was measured as larger than Earth.

Within 17 days of the planet's visual discovery in 1846 by Johann Gottfried Galle based on the mathematical

predictions of Urbain Le Verrier, the planet's first moon was also identified, by William Lassell. Triton is the largest of Neptune's 14 known moons, the latest, S/2004 N 1 confirmed by observations by the Hubble Space Telescope in 2013. Six small rings encircle the planet primarily formed of dust which, on occasion, appears to have defied the accepted norm and has clumped together to form arcs.

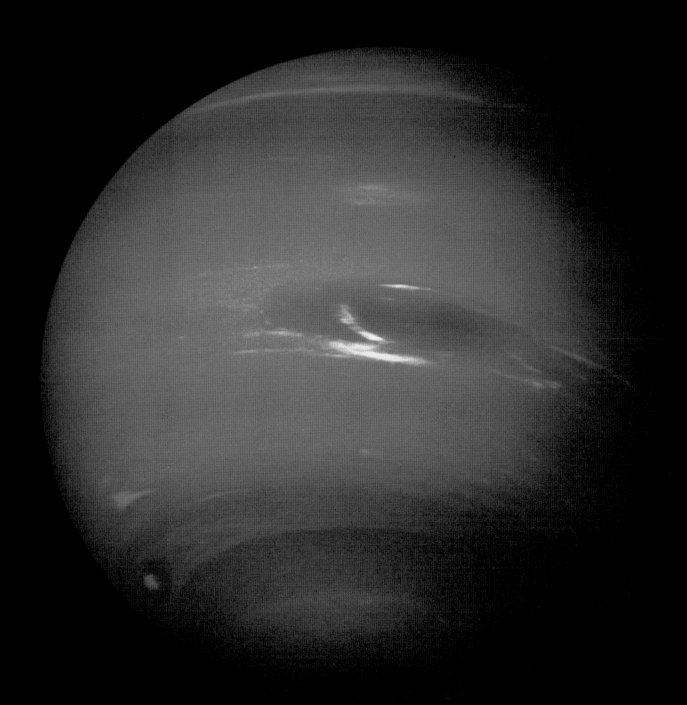

ABOVE: Neptune as imaged by Voyager 2 from a distance of 4.4 million miles, depicts the Great Dark Spot in its central region and a number of cloud systems and storms whipped up by interactions and powerful winds in the planet's atmosphere. Some of Neptune's winds have been measured racing at more than three times the speed of winds on Jupiter.

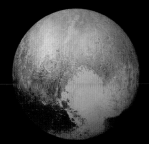

Perched on the edge of the Solar System, some 39 times farther away from the Sun than Earth, Pluto takes 248 years to orbit our star and 153 hours to complete a revolution about its axis.

PLUTO

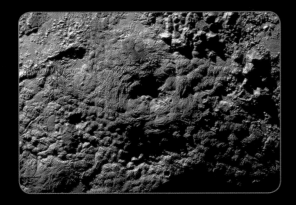

THIS PAGE, TOP: Output from New Horizons' Long Range Reconnaissance Imager (LORRI) was combined with color data from the spacecraft's Ralph instrument to create this global view of Pluto from a distance of approximately 280,000 miles away.

THIS PAGE, ABOVE: The 90 mile wide Wright Mons on the surface of Pluto close to the large plain, Sputnik Planum, is one of two sites identified by planetary scientists as a potential result of cryovolcanic activity.

It receives just 1/900th of the Sun's energy we experience on Earth. Electrically-charged particles from the Sun strip away some of Pluto's ultra-thin atmosphere giving it a long tail, a little like a comet's.

Pluto was considered the Solar System's ninth planet from its discovery by 24 year old Clyde Tombaugh at Arizona's Lowell Observatory in 1930 until 2006 when it was downgraded by the International Astronomical Union to the status of a dwarf planet. With a diameter of 1,430 miles, it is smaller than the Moon but the largest of the dwarf planets. Following his death in 1997, a small portion of Tombaugh's ashes were carried in the New Horizons spacecraft which became the first probe to reach Pluto in 2015. It discovered a more varied world than expected including rocky highlands one and a half miles higher than the surrounding icy plains, large cliffs, craters and glaciers. Tombaugh is commemorated by the naming of Tombaugh Regio - a 990 mile-wide, heart-shaped glacier of frozen nitrogen just north of the dwarf planet's equator which covers an area equivalent to the size of Texas and Oklahoma. New Horizons also revealed several potential ice volcanoes including Wright Mons. Named after the aviation pioneers, the Wright brothers, it is approximately 90 miles across and 2.5 miles high.

Pluto is orbited by five moons: Kerberos, Nix, Styx, Hydra, and Charon, the latter far and away the largest at around half the size of its parent planet. They were all named after mythical figures of the underworld, a convention started by an eleven year old British schoolgirl, Venetia Burney, who named Pluto in 1930 after the Roman god of the underworld.

THIS PAGE: A composite of enhanced color images of Pluto (lower right) and its major moon, Charon taken by NASA's New Horizons spacecraft in July, 2015, highlights the significant differences in terrain and surface properties between the dwarf planet and its moon. New Horizons got to within 17,000 miles of Charon. Pluto and its moon orbit each other every 6.387 days and are gravitationally locked to one another, so both keep the same side facing each other.

A star or stellar cluster is a group of stars that share a common origin and are gravitationally bound for some period, not necessarily forever.

STAR CLUSTERS

Newborn stars, hidden behind thick dust, are revealed in this image of a section of the Christmas Tree cluster (also known as NGC 2264) from the NASA Spitzer Space Telescope. The cluster lies around 2,700 light years away in the constellation Monoceros.

Astronomers find them useful in mapping and modelling stellar ages and evolution. Clusters come in two distinct types: globular and open clusters. Stars in an open cluster have a common origin, forming from the same initial giant molecular cloud. These clusters frequently contain stars numbered in hundreds although some contain mere dozens and others multiples of thousands. Stars in an open cluster are only loosely bound by gravity and may break up and spread apart as they orbit their galaxy's core. They tend to comprise relatively young stars; the Alpha Persei cluster, for example, is just 50 million years old whilst the most famous open cluster of all, the Pleiades is thought to be around twice that age.

Globular clusters are predominantly spherically-shaped groupings of old to extremely old stars, typically numbering between 100,000 and one million objects. The 60-light-year-wide Messier 28 in the constellation Sagittarius,

for example, is thought to be around 12 billion years old. Messier 30 in the constellation Capricornus comprises as many as 200,000 stars packed into a 90 light year diameter and is older, with an age estimated at 12.93 billion years. There is little free dust or gas to be found in globular clusters so new star formation is rare and many of their stellar population have already completed their main sequence. There are more than 150 globular clusters in the Milky Way, most occupying space in the halo surrounding the galaxy.

RIGHT PAGE: Stars dazzle in the massive star forming region 30 Doradus, some 170,000 light years from Earth. In this Hubble Wide Field Camera 3 image, taken in 2009, the blue light comes from the hottest, most massive stars.

THIS PAGE, TOP: This Hubble Picture of the Week shows Messier 28, a globular cluster in the constellation of Sagittarius (The Archer), in jewel-bright detail. It is about 18 000 light-years away from Earth.

Located below the clear skies of La Palma island in the Canary Islands, the Gran Telescopio Canarias is perched at an elevation of 7,438 feet on the summit of Roque de los Muchachos, one of 16 telescopes located there.

GRAN TELESCOPIO CANARIAS

Inside the dome reveals part of the GTC's giant steel telescope mount which holds the mirrors and allows rotational movements of the telescope along a horizontal and vertical axis. This movement needs to be extremely precise for the telescope to observe at high resolutions in both optical and infrared light ranges.

Following tests with part of its mirror in place, the telescope opened officially in 2009 after nine years of construction at an estimated cost of $180 million and became the world's single large reflecting telescope. Housed inside a 147-foot-tall dome, the GTC or 'Gran Tecan' boasts a 31.4-foot diameter primary mirror which possesses 65 square feet more light collecting area than any other land-based telescope. It's a far cry from the seven inch diameter reflector that English scientist Robert Hooke was said to wield in the 17th Century; the GTC has a diameter more than four times the size of the fabled Hooker Telescope at California's Mount Wilson Observatory used by Fritz Zwicky in the 1930s to find evidence of dark matter and by Edwin Hubble to determine the expansion of the Universe.

The GTC's 36 hexagonal segments of its main mirror are made of 3.15 inch-thick Zerodur glass-ceramic. They're so precisely formed and smoothed that if an individual pane were the size of the Iberian Peninsula (comprising all of Spain and Portugal), then no part of mirror would rise more than 0.039 inches high. These segments are all aligned with unerring accuracy by electric motors to give unrivalled precision. A partnership between Spain, Mexico and the University of Florida, the GTC has been already used to detect microquasars, detect new planetary systems around distant stars and is said to be able to detect the flickering flame of a single, small candle from 20,000 miles away.

The protective dome of the Gran
Telescopio Canarias. Inside the dome,
the movable structure of the GTC
weighs approximately 440 tons.

Historically described by the Ancient Greeks as "hairy stars" and now often categorised as "cosmic snowballs," these small balls of ice, dust and rock mostly orbit the Sun on long elliptical trajectories that occasionally bring them into relatively close view.

COMETS

Comets contain small solid nuclei typically between 1,000 feet and 25 miles in diameter. Occasionally, a larger nucleus is detected such as the 35-40 mile wide Comet Hale-Bopp, discovered jointly in 1995 by Alan Hale in New Mexico and Thomas Bopp in Arizona.

Beyond Jupiter, a comet remains cold and effectively dormant. As it gets closer to the Sun, though, some dust and ice form a hazy cloud around the nucleus and called a coma which can be hundreds of times larger than the nucleus. Solar radiation penetrates the dusty coma and causes ice in the comet's nucleus to sublimate and turn directly from a solid to a gas. The radiation pushes dust particles away from the coma, forming a dust tail, while charged particles from the sun convert some of the comet's gases into ions, forming a second stream of matter trailing behind the comet, known as an ion tail. Tails can extend huge distances. In 1996, a Japanese amateur astronomer, Yuji Hyakutake, discovered a comet (labelled C/1996 B2 or Comet Hyakutake). The Ulysses space probe inadvertently passed through the comet's tail and measured it at 360 million miles in length – close to four times the average distance between the Earth and Sun.

More than 6,300 comets have been identified, 32 of which were identified by Carolyn S. Shoemaker – the most by any one astronomer. What makes the feat more remarkable is that this New Mexico-born mother of three only took up astronomy seriously at the age of 51. Among her joint discoveries was the famous Shoemaker-Levy 9 comet which suffered a spectacular ending in 1994, ripped apart by Jupiter's gravitational force with the fragmented parts crashing into the planet.

Hale-Bopp comet.

ABOVE: Comet Lovejoy is visible near Earth's horizon in this nighttime image photographed by NASA astronaut Dan Burbank, Expedition 30 commander, onboard the International Space Station on Dec. 22, 2011.

ABOVE RIGHT: Comet 67P/Churyumov-Gerasimenko was taken by the Philae lander of the European Space Agency Rosetta mission during Philae descent toward the comet on Nov. 12, 2014 from a distance of approximate two miles three kilometers.

Comets are classified as either long-or short-period comets based on the time it takes them to complete their orbit. Short-period comets take 200 years or less to complete orbit. Some have exceptionally short-periods, such as Comet Wild 2, which takes just 6.39 years, and Encke's Comet, which orbits every 3.3 years. Many long-period comets only make a once-in-a-lifetime pass by of Earth giving astronomers a single chance to study them intensely. Comet C/2006 P1 McNaught has an orbital period of approximately 92,000 years so when it swung by in 2006 and 2007 and proved the brightest comet viewed for 40 years (bright enough to be viewed in daylight despite possessing a small 6-12 mile wide nucleus) it was understandably headline news.

Records of observations of Halley's Comet extend back over 2,000 years to China during the Han Dynasty and were written down in Sima Quan's Records of the Grand Historian. The comet even features on part of the world-famous Bayeux Tapestry, created to commemorate the successful Norman invasion of England in 1066CE. An

English astronomer, Edmund Halley, developed the ideas of celebrated scientist Sir Isaac Newton that comets orbited on elongated elliptical paths and published his calculations of the paths of 24 comets and their orbits in 1705. The orbits of three bright comets observed in 1531, 1607, and 1682 were so similar Halley deduced they were the same object returning to perihelion (its closest approach to the Sun) at an average interval of 76 years.

When Halley's comet returned in 1986, a fleet of spacecraft were ready to study it. These included two from Japan (Suisei and Sakigake) and Vega 1 and Vega 2 from the Soviet Union. The European Space Agency's Giotto probe travelled the closest, to within 373 miles of the comet's nucleus. Since that time, a number of missions have been made to comets including NASA's Deep Impact to Comet Tempel 1 and the ESA's Rosetta mission which orbited and studied Comet 67P/Churyumov–Gerasimenko between 2014 and 2016 and landed the Philae space probe on the comet's surface.

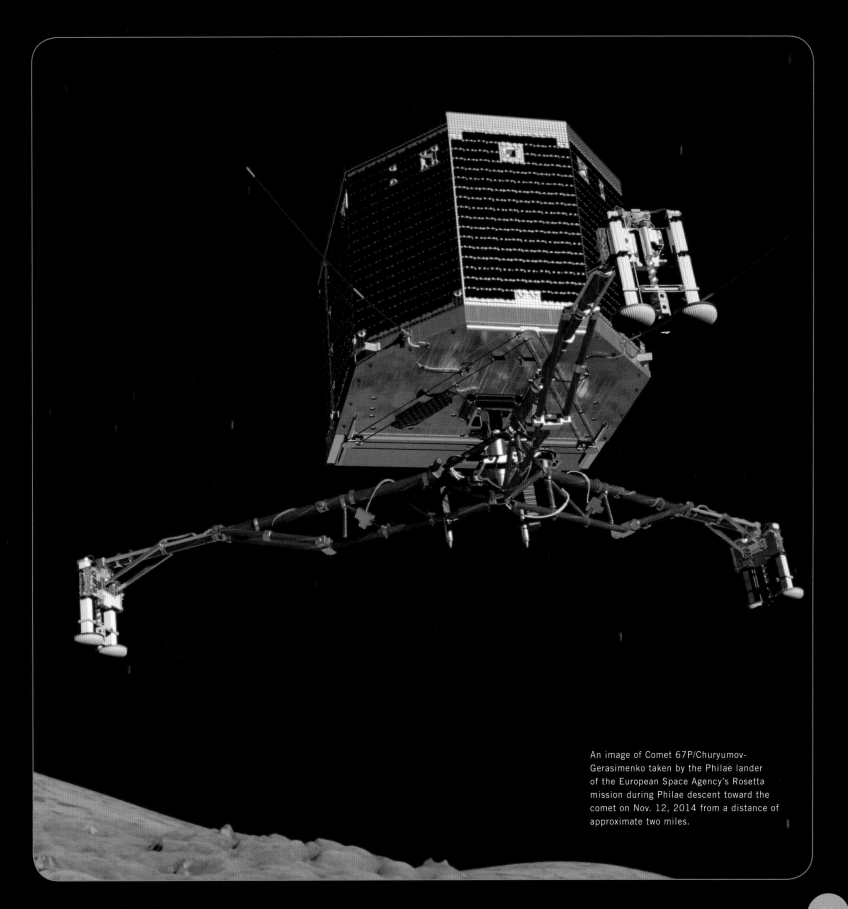

An image of Comet 67P/Churyumov-Gerasimenko taken by the Philae lander of the European Space Agency's Rosetta mission during Philae descent toward the comet on Nov. 12, 2014 from a distance of approximate two miles.

Extending around the fringes of the Solar System past Neptune's orbit, the Kuiper Belt begins some 3.5 billion miles from the Sun.

KUIPER BELT

ABOVE: An artist's visualization of the Kuiper Belt object, 90377 Sedna. This dwarf planet is 600-660 miles in diameter and possesses an extremely elongated elliptical orbit which at perihelion takes it as close as 76 astronomical units (AU) from the Sun but at aphelion sends it as far as 937AU away or more than 30 times Neptune's average distance from the Sun.

This cold, dark third zone of the Solar System is thought to contain millions of small, rocky and icy bodies, remnants and leftovers from the formation of the Solar System over four and a half billion years ago. Some bodies in the belt, including Pluto and Eris, are large enough to be classed as dwarf planets. Two further dwarf planets, Haumea and Makemake, were discovered in the Kuiper Belt in 2008, whilst Sedna was discovered in 2003 and named after the Inuit goddess of the ocean. This object, at least 600 miles in diameter, has a highly distant and eccentric orbit stretching at aphelion, more than 900 AU from the Sun and taking 11,400 years to complete a single orbit.

The first spacecraft to enter the Kuiper Belt region was NASA's Pioneer 10 in 1983 but it was New Horizons in 2015 that was the first to examine a KBO (Kuiper Belt Object) when the probe reached Pluto. Four years later, it performed a fly-by of a second KBO – an unusual pair of planetisimals, one 12 miles wide, the other 9 miles, bonded together and called 2014 MU69 or better known by the nickname, Ultima Thule.

Beyond the Kuiper Belt is thought to lie the theorized Oort Cloud. Named after Dutch astronomer Jan Oort, this shell or bubble extends to the outermost boundaries of the Solar System from at least 1,000 Astronomical Units (AU) from the Sun out as potentially far as 100,000 AU. Even at its current rate of travel of one million miles per day, the Voyager 1 spacecraft would take 300 years to reach the Oort Cloud's inner boundary. The cloud is thought to contain many trillions of icy objects and is the source of most long-period comets.

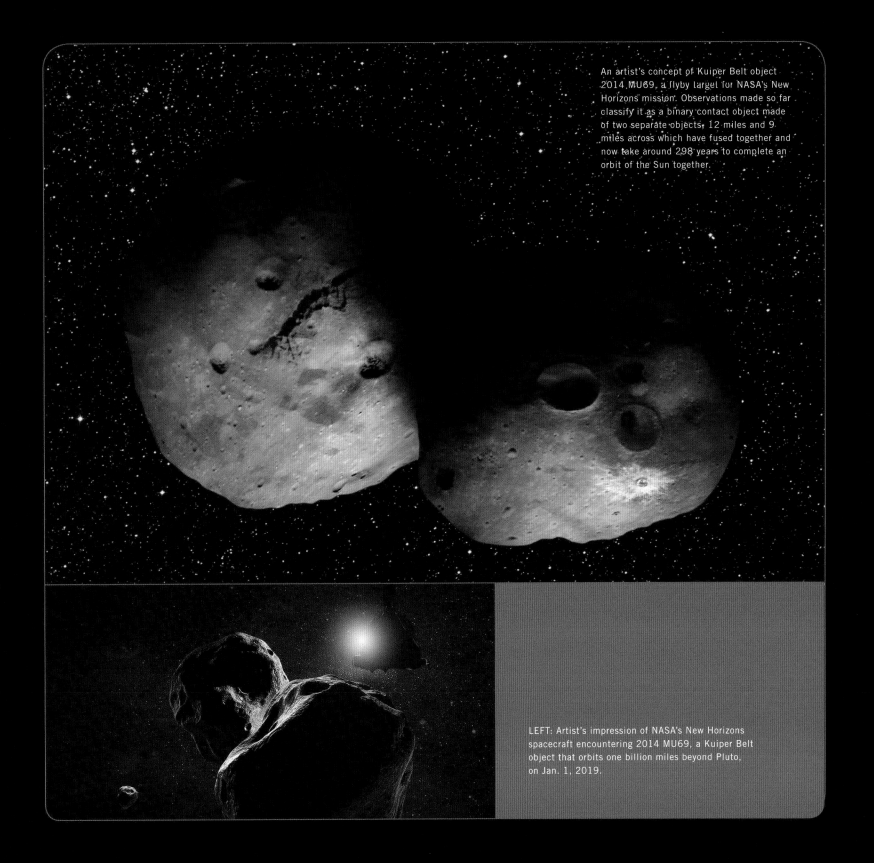

An artist's concept of Kuiper Belt object 2014 MU69, a flyby target for NASA's New Horizons mission. Observations made so far classify it as a binary contact object made of two separate objects, 12 miles and 9 miles across which have fused together and now take around 298 years to complete an orbit of the Sun together.

LEFT: Artist's impression of NASA's New Horizons spacecraft encountering 2014 MU69, a Kuiper Belt object that orbits one billion miles beyond Pluto, on Jan. 1, 2019.

OUTSIDE THE
SOLAR SYSTEM

  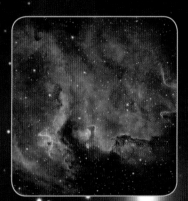 

Stars are born, shine brightly, and eventually die – a
life cycle played out over millions or billions of years.

STARS

These massive balls of hot gas held
together by their own gravity begin life
in nebulas where collapsing molecular
clouds draw gas and dust together
with accretion building mass and
hence increasing gravitational force
and the ability to draw more matter
in. Temperatures and pressures rise,
especially in the core, and the mass may
start spinning to form a protostar. As
the protostar gets bigger and hotter,
pressure and temperatures continue
to increase in its core until nuclear
fusion reactions begin. In a Sun-like
star, a temperature of some 27 million
degrees Fahrenheit is required for
fusion reactions where millions of tons
of hydrogen gas each second act as fuel
driving proton-proton chain reactions
which create helium, excess hydrogen
nuclei, and phenomenal amounts of
energy. Whilst fusing hydrogen, stars
are said to be undergoing their main
sequence. Around 90 per cent of all
stars in the Milky Way are in their main
sequence.

Mass and duration are inextricably
linked during a star's main sequence.
Stars around the same size as the
Sun can expect a 10-billion-year-long
main sequence, but duration decreases
markedly for larger mass stars.
Those with masses of approximately
three times the Sun's may have a
main sequence of just 370 million
years. During their main sequence,
stars undergo a balancing of forces;
the inward force of gravity balanced
by the outwards force generated by
fusion reactions and the resulting gas
pressure and radiation.

THIS PAGE: Sh2-284, a star-forming cloud of dust and gas stands out vividly in this image taken by NASA's Wide-field Infrared Survey Explorer. Inside the nebula, an open star cluster called Dolidze 25 is emitting vast amounts of radiation in all directions.

The eerie glow of debris and gas from a previously exploded supernova is depicted in this Hubble Space Telescope image of the famous Crab Nebula.

Astronomers classify stars in multiple ways. In the 1910s, Danish astronomer Ejnar Hertzsprung and American Henry Norris Russell devised the Hertzsprung-Russell (H-R) diagram depicting the relationships between a star's temperature and its colour and luminosity – a useful tool for grouping stars and plotting stellar evolutionary life cycles. Main sequence stars vary widely in their surface temperatures. At around 9,900 degrees Fahrenheit, the Sun is exceeded by Sirius (18,000 degrees) – the brightest star in the night sky from Earth, and eclipsed by Theta-1 Orionis C – a star in the Orion Nebula (~70,000 degrees).

The smallest stars, known as red dwarfs, may contain as little as 7.5 per cent the mass of the Sun and emit only 0.01 per cent as much energy. Despite their low luminosity and temperature (below 6740 degrees Fahrenheit), they are thought to be the most populous star in space, at least within the observable reaches of current astronomy. In contrast, far rarer supergiants and hypergiants are many times the size of the Sun. In 2014, the European Southern Observatory's Very Large Telescope revealed a potential yellow hypergiant, HD 119796A (also known as V766 Centauri) 12,000 light years away with a diameter at least 1300 times greater than the Sun.

End of Life
Stars eventually run out of hydrogen. This can take millions or billions of years depending on their mass. When small stars exhaust their fuel, they slowly dim in brightness, eventually cooling and fading away. Dying stars of similar mass to the Sun swell to become a red giant, expel outer layers to form a planetary nebula, and then contract to become a small, dense white dwarf which cools over time. The endgame for stars possessing many times the mass of the Sun can be considerably more violent. They swell to supergiants, fuse helium, then successively heavier chemical elements until the core can no longer support the expanding star. This may lead to a powerful implosion and collapse followed by a devastating rebound that rips through the star, blowing it apart in a type II supernova.

A star in its death throes lasting thousands of years ejects clouds of gas after strong stellar winds push away the star's outer layers. The gas is are illuminated by the dying star's still-energetic core. This particular planetary nebula, called NGC 6565, is some 14,000 light years away in the constellation Sagittarius.

Visible from most parts of the globe, the Pleiades (also known as Messier 45) is one of the brightest star clusters viewable from Earth.

THE PLEIADES

Lying in the Taurus constellation some 444 light years from Earth and spanning an area over 80 light years across, the Pleiades cluster contains at least 1,000 observed individual stars and, in reality, many, many more. It may also contain large numbers of brown dwarfs – failed star bodies - making up as much of a quarter of the population but a mere two per cent of the cluster's mass. The Pleiades are most clearly seen from Earth during winter in the Northern

Hemisphere and summer in the Southern. Light from the stars reflects off passing interstellar clouds of gas and dust giving the stars and the cluster overall a hazy, blue-white appearance.

Studied since ancient times, the cluster gets its name from the seven daughters of Atlas and Pleione in Greek mythology and were first spotted by telescope by the pioneering astronomer, Galileo Galilei. The central core, with a radius of around seven light years, is home

to nine exceptionally bright blue stars including Maia, Electra, and Taygeta and collectively known as the Seven Sisters. The largest of their number, Alcyone (Eta Tauri), was thought to possess approximately 9.2 times the radius of the Sun and shines some 2,030 times more brightly. In recent times, it has been ascertained that Alcyone is actually a multiple star system comprising at least three objects.

The Pleiades as viewed by NASA's Wide-field Infrared Survey Explorer (WISE) space telescope. First catalogued as M45 by Charles Messier, the Pleiades are loosely bound gravitationally and this open cluster is expected to disperse over a period of around 350 million years.

LEFT: Imaged by NASA's Spitzer Space Telescope in the infrared part of the spectrum, the Pleiades shine through the surrounding swirling clouds of dust and gas.

Variable stars experience fluctuations in their absolute luminosity (brightness) from as little as a thousandth of a magnitude to many magnitudes.

VARIABLE STARS

The periods that these changes occur over vary greatly as well; from fractions of a second to many years depending on the type of variable star, over 150,000 of which have already been catalogued since the first, Omicron Ceti (now called Mira), was observed pulsating over a 11 month cycle by German astronomer Johannes Holwarda in 1638.

Changes in a star's brightness can occur due to internal or external factors. Pulsating variables, for example, swell and shrink due to internal forces with the outward expansion of gases produced by core nuclear fusion battling for balance with the inwards pulling force of gravity.

Binary stars (typically two stars which orbit around their common centre of mass) are an example of variable stars due to external factors, one star eclipsed by the other then moving out of its way as they orbit. Algol is a classic example of an eclipsing binary.

In 1907, Henrietta Swan Leavitt, whilst working at Harvard College Observatory, discovered that Cepheid variable stars in the Small Magellanic Cloud pulsated at a rate which depended on their absolute magnitude. This enables astronomers to use Cepheid variables as 'standard candles' – a useful tool in determining distances. These stars can be observed and measured to a distance of about 20 million light years. The southern hemisphere Cepheid variable RS Puppis, for example, has a period of 41.5 days during which it varies in apparent magnitude from 7.67 to 6.52. A supergiant star, it has ten times the mass of the Sun and an average intrinsic brightness 15,000 times greater than our star, making it one of the brightest Cepheid variables in the Milky Way.

A Hubble Space Telescope image of the variable star, RS Puppis, shows it swaddled in a cocoon of reflective dust. According to NASA, the star is 200 times larger than the Sun and has approximately ten times the Sun's mass. Its average intrinsic brightness is 15,000 times greater than the Sun's luminosity. The nebula flickers in brightness as pulses of light from the star propagate outwards.

Discovered in 1677 by Edmond Halley (after whom the famous comet is named), Eta Carinae is a large and wildly unstable star system around 7,500 light years from us in the constellation Carina.

ETA CARINAE

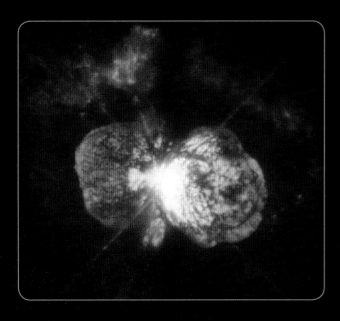

LEFT: This Hubble Space Telescope image depicts the Homonculus Nebula that surrounds Eta Carinae - the result of a colossal outburst of material viewed from Earth during the 1840s. The Homunculus consists of two giant lobes, referred to as NW and SE based on their orientation as seen from Earth. Holes, trenches and dust skirts in this emission nebula have been detected.

The two stars which orbit around their common center of gravity once every 5.5 years, have a combined mass of around 120-130 times that of the Sun but shine with a brilliant intensity, boasting a combined luminosity of up to five million Suns. Eta Carinae's brightness increased more than a hundredfold during a period known as the, "Great Eruption" (1838–43). For a short spell in March 1843, it became the second brightest star in the night sky after Sirius (which was almost 1,000 times closer to Earth) before

dimming and since, has continued to fluctuate greatly.

The Great Eruption was so named after the vast quantities of gas and dust shed by the star system - enough matter it is estimated to form at least ten Suns. The ejected material formed the Homonculus Nebula with its distinctive lobes which surrounds the stars today and is what is viewed in this image, obscuring its parent star from view. The nebula has been studied extensively and its colliding

stellar winds emanating from both stars which get as close as 140 million miles apart (roughly the average distance Mars lies away from the Sun) are believed to lead to temperatures exceeding 70 million degrees Fahrenheit. It is thought that Eta Carinae is heading for an explosive ending sometime within the next one or two million years with the larger, more unstable of its stars expected to go supernova.

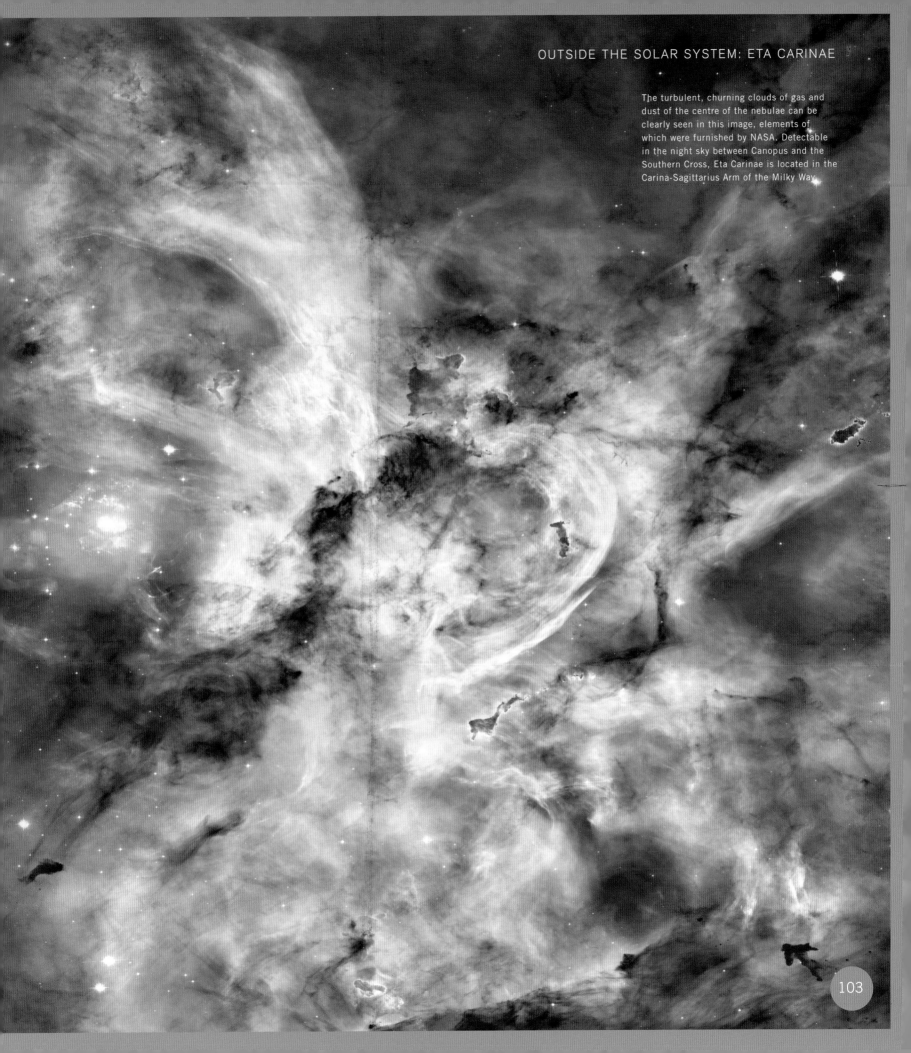

The turbulent, churning clouds of gas and dust of the centre of the nebulae can be clearly seen in this image, elements of which were furnished by NASA. Detectable in the night sky between Canopus and the Southern Cross, Eta Carinae is located in the Carina-Sagittarius Arm of the Milky Way.

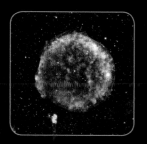

Supernovae tear stars apart with searing intensity and unimaginable force.

SUPERNOVA REMNANTS

The energy these star-shredding explosions produce can exceed the entire output of a medium-sized galaxy containing billions of stars. Supernovae occur relatively rarely in an individual galaxy, typically at a rate of once or twice per century, and astronomers, amateur and professional, scour the skies, hoping to detect one. A ten year old Canadian schoolgirl, Kathryn Aurora Gray, holds the title of youngest supernova hunter, when she discovered a SN 2010lt in the galaxy, UGC 3378.

There are two main classes. Type Ia supernovae occur in binary star systems featuring a white dwarf. These small, ultra-dense stars - a teaspoon of which may weigh as much as five tons according to the Hubblesite - may have enough gravitational pull to gather hot gaseous matter from their companion star if it orbits close enough. If the white dwarf increases its mass to reach a level equal to 1.4 solar masses and known as the Chandrasekhar Limit (after the Indian-American astrophysicist Subrahmanyan Chandrasekhar), the star will collapse catastrophically before uncontrollable, runaway fusion of carbon and oxygen effectively detonates the star, ejecting material at phenomenal speeds. The resulting light from the explosion of a Type Ia supernova is rated at five billion times brighter than the Sun.

Type II supernovae see the demise of massive stars, eight times or more massive than the Sun which swell, use increasingly heavy elements as fuel for fusion reactions in their core and then when the fuel is exhausted, lack of outward pressure causes the star to collapse in on itself sharply. The star's matter rebounds outwards suddenly and with spectacular violence, propelled by a massive neutrino outburst and shock waves, tearing the star apart and scattering its matter outwards into space.

ABOVE: The Tycho supernova remnant as imaged using both infrared and X-ray observations from different space telescopes.

The Soul Nebula, a large
star-forming region in the
constellation, Cassiopeia.

"Precise information about the frequency of the different supernova types in our galaxy and its surroundings will shed light on the star-formation history and chemical evolution of the local group of galaxies."

Dr Andrea Pastorello, supernova researcher, Padova Astronomical.

This image from Wide-field Infrared Survey Explorer (WISE) depicts the supernova remnant IC 443, also known as the Jellyfish Nebula. The result of a Type II supernova, IC 443 is heavily studied for its interactions with the molecular clouds surrounding it.

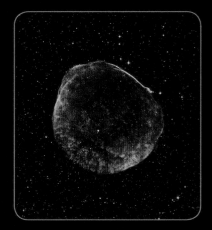 

ABOVE LEFT: The remains of the SN1006 supernova that were first observed just over a millennia ago, span a region of space approximately 60 light years in width. It is almost invisible in the optical part of the spectrum so is imaged using X-ray and radio telescopes.

ABOVE RIGHT: Imaged by the Chandra X-Ray Observatory and infrared observations from the ground, the W49B supernova remnant lies around 26,000 light years away. As seen from Earth, it is approximately one thousand years old and contains iron, manganese and sulfur amongst other, heavier elements, within its material.

A supernova's incredibly intense energy burst is a temporary phenomenon. One of the brightest known supernova (SN 1006) was observed at the end of April 1006CE by astronomers as far afield as China, Egypt and Switzerland. Despite its location 7000 light years from Earth, this supernova was reported as shining brighter than Venus and still viewable in the night sky three years after the initial observation, according to official Chinese Song Dynasty records.

A supernova destroys a star as we know it but they do leave behind significant remains. The core of the original star may form a small but exceptionally dense neutron star whilst the scattered material, now rich in heavier elements due to fusion processes during the explosion, may form a supernova remnant. SN 1006's remnant wasn't discovered until 1965, by Australian astronomers Frank Gardner and Doug Milne using the Parkes radio telescope.

Supernova remnants continue to expand often with considerable velocity centuries after the explosion that caused them. Puppis A, the remnant from a supernova that light reached Earth from around 3,700 years ago is now estimated to have expanded to a width of 100 light years. When a type Ia supernova, SN 1572 occurred in the constellation, Cassiopeia in 1572, it was said to be so bright, it could be seen during the day and entranced astronomers including the Dane, Tycho Brahe who entitled his book, De Nova Stella after the event. The Chandra X-ray Observatory measured the remnant, now commonly known as Tycho's Supernova Remnant, in 2000 and 2015 and detected that during that short time period, it had expanded by an estimated 100 billion miles.

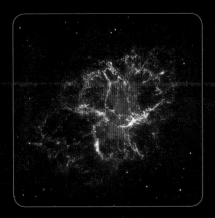

In 1054CE, astronomers in both the Middle East and China reported a sudden appearance of a new star in the constellation Taurus.

THE CRAB NEBULA

Its luminosity was said to be so great that it could be observed during daylight for 23 days. This "guest star" was in fact a violent supernova explosion occurring around 6,500 light years away. The remains of this supernova, rediscovered using telescopes by English doctor John Bevis in 1731, are known as the Crab and also NGC 1952 and Messier 1, and continue to expand at an extremely rapid rate. The nebula is already more than 11 light years across and its outer layers and filaments of hot gas are moving outwards at a rate of approximately three million miles per hour. Measurements taken by the Spitzer Space Telescope indicate that gases are still glowing at temperatures of 20,000 to 32,000 degrees Fahrenheit 1,000 years after the supernova occurred.

Differing chemical elements within the expanding gas help provide the varying colors when a composite image of the nebula (such as here) is made. These include hydrogen (orange), nitrogen (red), sulfur (pink), and oxygen (green).

At the very centre of the nebula lies the end game for the once massive star with a mass 8-15 times the Sun, which was torn apart in the explosion. It is the Crab Pulsar (also known as NP0532), this rapidly spinning dense neutron star completes 30 rotations per second (one rotation every 33 milliseconds) and is gauged as being 100,000 times more energetic than the Sun. According to NASA, the star is no bigger than a town on Earth yet has as much mass as the Sun.

ABOVE: The Herschel Space Observatory and the Hubble combine to produce this composite view of the iconic Crab Nebula in the constellation, Taurus.

ABOVE: A mosaic of 24 Hubble images taken in 1999 and 2000 form this stunning portrayal of the Crab Nebula and its expanding, complex structure of filaments. Blue in the filaments indicates the presence of oxygen whilst green depicts the presence of ionized sulfur.

Just under 350 years ago, light from this spectacular supernova remnant first reached Earth, not that we are sure if anyone was able to spot it as it is stronger in wavelengths other than visible light and may have been obscured by interstellar dust.

CASSIOPEIA A

When it was finally discovered in 1948 by radio astronomers Martin Ryle and Francis Graham-Smith, it was spotted as a radio source and only identified visually two years later.

Equipped with powerful instruments, orbiting space observatories such as Chandra, Spitzer and the Hubble are now able to bring this celestial wonder to life. What their imaging depicts amongst the knots and filaments of gas and dust are the rapidly-expanding remains of the end of a star, which is between 10 and 30 light years wide and with some of the material travelling at an almost incomprehensible 8.9 million to 13.4 million mph. Located in the Cassiopeia constellation around 11,000 light years away, the expanding cloud reaches temperatures of 50 million degrees Fahrenheit, conveying an idea of the phenomenal energy involved in a supernova explosion.

Supernovas are considered a key distributor of chemical elements throughout the Universe. Studies of Cassiopeia A reveal the supernova remnant contains hydrogen, carbon, nitrogen, phosphorous and oxygen – all the key chemical elements required to make DNA. Scientific instruments has measured the prodigious quantities the remnant has already of certain elements including enough sulfur to equal the mass of 10,000 Earths and around 70,000 Earth masses of iron.

TOP LEFT: The magenta burst in Cassiopeia A, as identified, by the Fermi Gamma-ray Space Telescope, indicates the location within the supernova remnant where powerful radiation is being emitted more than one billion times more energetic than visible light.

ABOVE: The shimmering ghostly haze of Cassiopeia A's shells of gas and debris form
a remarkable image of the remains of a once massive star that went supernova.

A magnificent image requires a suitable lofty name which NASA supplied to this stunning image of part of the Eagle Nebula.

PILLARS OF CREATION

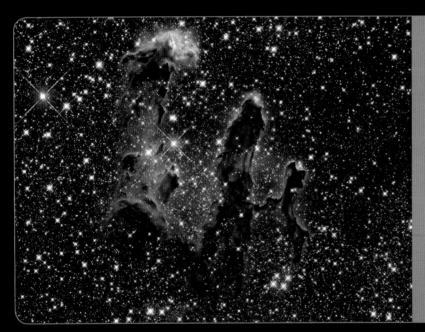

HUBBLE GOES HIGH DEF TO REVISIT THE ICONIC 'PILLARS OF CREATION'

This NASA Hubble Space Telescope image, released in 2015, offers a contrasting view of the Pillars of Creation compared to the classic portrayal. Taken in near-infrared light, which penetrates much of the gas and dust in the region, the pillars are transformed into eerie, wispy silhouettes, viewed against a background of large numbers of stars. The ghostly haze bordering the dense edges of the pillars is material heating up and evaporating away. The cause is surmised to be intense radiation from a cluster of young, massive stars nearby. The Pillars of Creation are just a small part of the Eagle Nebula which spans an area of space approximately 70 by 55 light years in size.

It was taken by the Hubble Space Telescope's Wide Field and Planetary Camera 2 on April 1, 1995 under the command of Arizona State University's Jeff Hester and Paul Scowen but is no April fool or hoax. The photo pieced together from 32 separate Hubble images depicts lofty clouds of interstellar gas, mostly hydrogen, and dust, around 6500-7000 light years away. The largest in shot on the left is around four light years in height. Its finger-like projection is estimated by astronomers to be larger than the entire Solar System.

Hot young stars in the pillars' vicinity bombard the columns of gas with intense radiation, causing ionization with atoms stripped of their electrons whilst stellar winds containing barrages of charged particles act as a cosmic sandblaster, eroding and blowing away the 'tops' of the pillars. The Pillars have been re-imaged on several occasions including by the European Space Agency's Herschel Space Observatory in 2010 and four years later, a revisit by the Hubble in advance of its 25th birthday celebrations, using its higher definition Wide Field 3

camera, fitted in 2009 to take both visible light and near-infrared images of this monumental structure. Comparing both images, astronomers were able to detect a jet-like phenomenon possibly spouted from a highly energetic young star, with the material travelling at speeds as high as 450,000 mph.

THIS PAGE: The classic 1995 image of the majestic Pillars of Creation, located in the Eagle Nebula which is found in the constellation Serpens.

Parts of the center of the Milky Way are hidden from Earth's viewpoint, blocked off by large sequences of overlapping molecular dust clouds located between the Solar System and the Sagittarius Arm, a minor spiral arm of our galaxy.

CYGNUS X

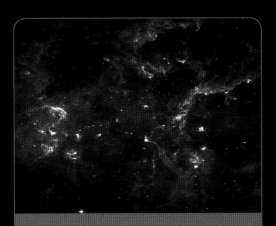

ABOVE: This infrared image depicts how intense radiation from Cygnus X's young groups of stars have heated and pushed gas away from the clusters, producing cavities of hot, lower-density gas. Ridges of denser gas mark the boundaries of the cavities whilst bright spots within these ridges show where stars are forming at present.

This phenomena is dubbed the Great Rift and Cygnus X lies behind it, making it hard to view in visible light. But when instruments capable of gathering in infra-red and other wavelengths study this region, they find a sensational large-scale star factory at work.

Described by NASA as, "a bubbling cauldron of star birth," this region of space, located in the Cygnus constellation and about 4,500 light years distant, is extremely active and turbulent, containing innumerable protostars across its 600 light year extent. Amongst the vast numbers of spidery filaments of gas and dust, dense knots and clumps of matter are heating up and collapsing down through gravity to form new massive stars. Infant stars are beginning the nuclear fusion reactions in their core to shine brightly. As these new, massive, stars power up, they blow away large bubbles and shells of hot, energetic gas which may be further energised by stellar winds and collisions which frequently trigger further clumping and collapse of matter nearby with the potential for yet more new stars to form.

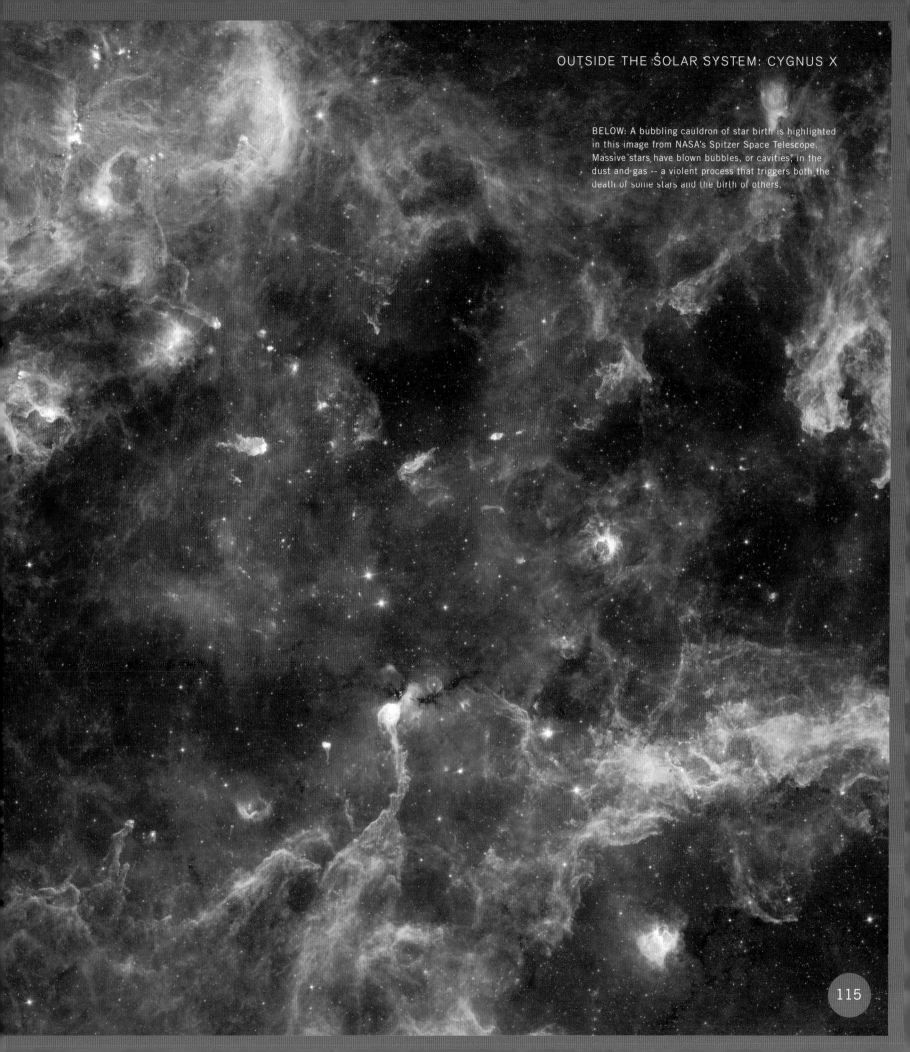

BELOW: A bubbling cauldron of star birth is highlighted in this image from NASA's Spitzer Space Telescope. Massive stars have blown bubbles, or cavities, in the dust and gas -- a violent process that triggers both the death of some stars and the birth of others.

This unusual and arresting planetary nebula in the Draco constellation lies between the Big and Little Dippers and was first spied by William Herschel in 1786.

CAT'S EYE NEBULA

Nebulae were not fully understood at that juncture and with the limited resolving power of optical telescopes of the time, many thought that nebulae were clusters of stars until ground-breaking work with a new device, a star spectrometer, was performed by William Huggins in 1864. In detecting and analysing the spectrum of light emitted from the Cat's Eye, Huggins excitedly concluded that, "The riddle of the nebulae was solved. The answer, which had come to us in the light itself, read: Not an aggregation of stars, but a luminous gas."

The Cat's Eye (also known as NGC 6543) is at least 3,300 light years away from us and has a core diameter of approximately one fifth of a light year. The Wolf-Rayet type star at its centre is shedding mass at a rate, measured in 2004 by Roger Wesson and X.W. Liu at a staggering 20 trillion tonnes per second. The nebula's resulting shells of dust and gas appear to be ejected from the core in a series of pulses or convulsions, creating a series of attractive, otherworldly balloon-like shrouds of matter; the Hubble Space Telescope has observed at least eleven differing shells as well as knots of gas, propelled and contorted by stellar winds. Some astronomers think that the winds are generated by an evolving binary star system at the core and its interactions.

BELOW: A bubbling cauldron of star birth is highlighted in this image from NASA's Spitzer Space Telescope. Massive stars have blown bubbles, or cavities, in the dust and gas -- a violent process that triggers both the death of some stars and the birth of others.

Nicknamed the "Eye of Sauron" from JRR Tolkien's Lord of the Rings fantasy novels, this planetary nebula is also catalogued as NGC 7293 and was discovered in the early 19th Century by German astronomer Karl Ludwig Harding.

HELIX NEBULA

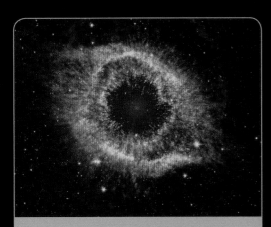

This infrared image from NASA Spitzer Space Telescope shows the Helix Nebula, a common target for amateur astro-photographers for its vivid colours and its eerie resemblance to a giant eye.

Recent measurements by the European Space Agency's Gaia space observatory indicate that the Helix Nebula, located in the Aquarius constellation, lies 655 light years away, making it the nearest planetary nebula to Earth. The sprawling oval-shaped nebula has an apparent magnitude of +7.4 making it bright for its kind, and at around two and a half to three light years in width, it is also the largest nebula in Earth's immediate vicinity.

The Helix Nebula has been studied extensively and observations have revealed many examples of cometary knots where faster moving gases tangle with slower moving, denser gases released at an earlier point. The nebula's outer layers of gas are moving away from the core, which contains a white dwarf too small to resolve with most current scientific instruments, at speeds of around 72,000 mph. This indicates that the nebula formed around 12,000 to 13,000 years ago. When imaged using infrared and ultraviolet instruments, the glow from a dusty disc can be identified encircling the central white dwarf star. Some astronomers postulate that this ring was formed from the remains of a ring of comets that once travelled in orbit around the star.

The Helix Nebula's outer layers
glow with the intense radiation
pumped out by the hot stellar
core as they gradually unravel out
into space.

One of the most-loved and impressive of all the hundreds of thousands of images taken by the Hubble Space Telescope is this small part of the Carina Nebula, around 7,500 light years away from Earth.

MYSTIC
MOUNTAIN

Taken by the newly installed Hubble's Wide Field Camera 3 in February 2010, Mystic Mountain looks like a fantastical wizard's hideaway or a bizarre science fiction landscape. The Hubble team described it as a "turbulent cosmic pinnacle [that] lies within a tempestuous stellar nursery."

The giant, craggy column of dust, hydrogen, oxygen and other gases measures about three light years tall or put another way, approximately 189,700 times the average distance between the Earth and the Sun. Yet, it is under assault from energetic, infant stars that exist both nearby and within the pillar. Their intense, scorching radiation and powerful stellar winds of charged particles are eroding the pillar relentlessly. Powerful streams and jets of gas launched by churning, rapidly spinning disks of matter that has been drawn in via accretion around infant stars, can be clearly viewed in the centre and peak of the gas pillar.

Chosen as the image to celebrate the Hubble's 20th anniversary of its time in space, Mystic Mountain will not last forever. In the battle between the energetic stars and this gargantuan column of dust and gas, the stars will win out. In the next few million years, Mystic Mountain will be gradually blown away and its matter distributed throughout a region of the universe.

TOP LEFT: This 2010 Hubble image of Mystic Mountain captures some of the intensely chaotic activity occurring within the gargantuan pillars of gas and dust attacked by intense radiation. By this time, the Hubble had taken 570,000 images of space.

Taken by the Hubble's Wide Field Camera 3 on 1-2 February 2010, this composite image corresponds to the glow of oxygen (blue), hydrogen and nitrogen (green), and sulphur (red).

As befits the word nebulous, many nebula appear amorphous and lacking structure or pattern to our eyes.

HORSEHEAD NEBULA

So, when a recognisable shape jumps out of us over the great reaches of space, we sit up and take notice. This is most definitely the case with the striking Horsehead Nebula, approximately 1,500 light years from Earth. This unusually-shaped cloud of dust and gas was first detected by Boston-born astronomer Edward Charles Pickering from a black and white photographic plate taken in 1889. It was re-discovered and revaluated by a gifted observational astronomer, Edward Emerson Barnard, who published the first description of the nebula in 1913. It made it into his pioneering 1919 catalogue of 182 dark nebulae as Barnard 33.

Between two and light years across, the nebula consists mostly of cold molecular hydrogen. The thickness of its clouds block light from the stars both contained within and behind, but it is illuminated by the ionizing gases of the bright emission nebula, IC 434. Astronomers also believe that the nebula may also be excited by the neighbouring bright star system, Sigma Orionis. A popular if challenging target for experienced astrophotographers, the nebula was the winner of a 1999 NASA Jet Propulsion Laboratory internet vote for a target for the Hubble Space Telescope to investigate. As Utah astronomer and blogger, Joe Bauman wrote, "Nothing in

astronomy captures the imagination like the Horsehead Nebula...a dark ghostly beast with a rolling mane, seen against a glowing pinkish curtain...for many it's the loveliest and most dramatic thing up there."

TOP LEFT: This ground-based image was taken by the NSF's 0.9-meter telescope on Kitt Peak using the NOAO Mosaic camera. Many stars in the lower half of the image are obscured by dark clouds of hydrogen gas.

THIS PAGE: Backlit wisps of matter along the Horsehead Nebula's upper ridge are illuminated by Sigma Orionis, a young five-star system just off the top of this image from the Hubble Space Telescope.

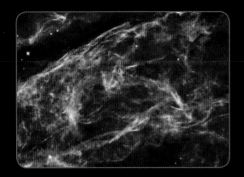

The visible part of the Cygnus Loop, many prefer the name, Veil Nebula to describe this supernova remnant's delicate, draped filament structures.

THE VEIL NEBULA

It was first sighted by the British astronomer William Herschel in 1784 and comprises the remains of a star 20 times the mass of the Sun which went supernova between 7000 and 10,000 years ago. At approximately 110 light years across, it is one of the largest supernova remnants to be studied in the night sky spanning a width of six full moons as seen from Earth but only appearing in faint outline to all but the most powerful of telescopes.

Located in the constellation Cygnus and lying roughly 2100 light years away, the remnant's cloud of hot gas is still moving rapidly at speeds of up to 370,000mph and as it collides with interstellar gas clouds, temperatures rises by millions of degrees Fahrenheit. According to Hubble Space Telescope observations, shock waves still race through parts of the structure at velocities close to one million miles per hour. The remnant emits radio waves and X-rays and its brightest regions are catalogued as NGC 6992 and

NGC 6995 and together known as the Eastern Veil. Further observations by the Hubble Space Telescope have revealed the presence of hydrogen (shown in the image in red) as well as heavier chemical elements including oxygen and sulfur, highlighting supernova remnants' role in disseminating heavy elements through space.

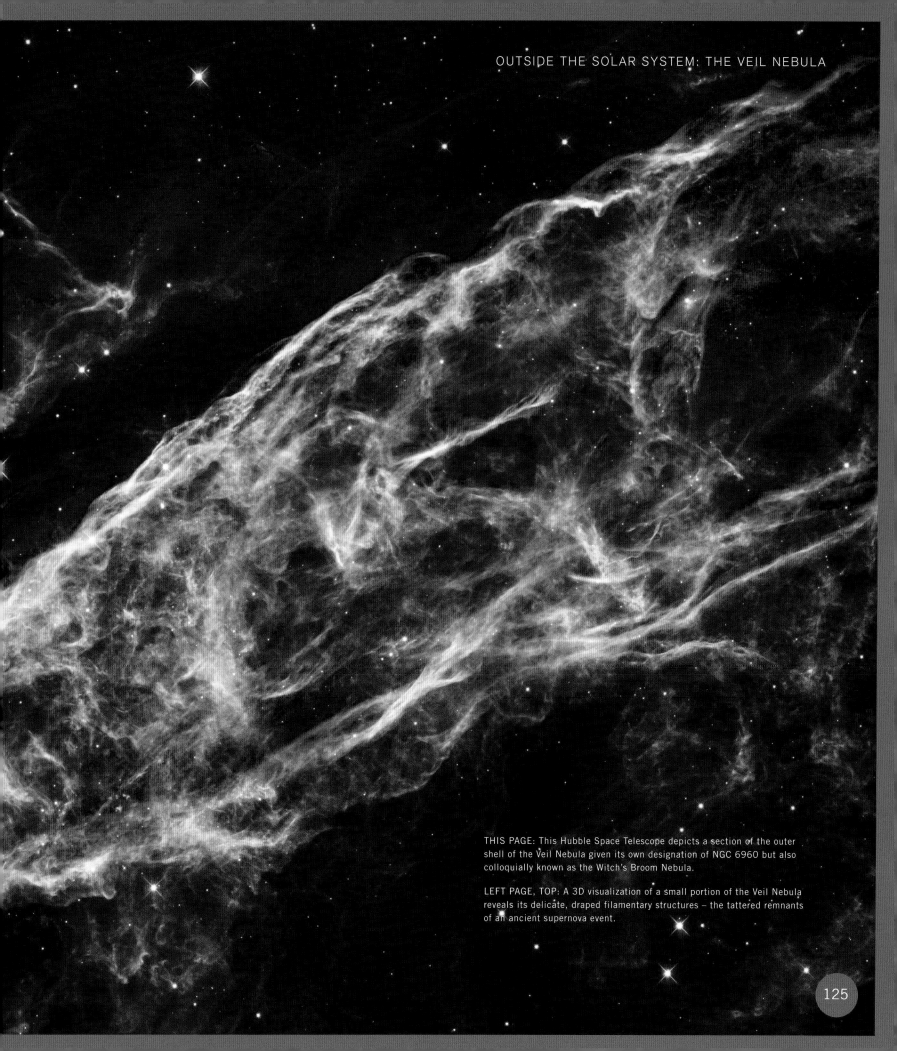

THIS PAGE: This Hubble Space Telescope depicts a section of the outer shell of the Veil Nebula given its own designation of NGC 6960 but also colloquially known as the Witch's Broom Nebula.

LEFT PAGE, TOP: A 3D visualization of a small portion of the Veil Nebula reveals its delicate, draped filamentary structures – the tattered remnants of an ancient supernova event.

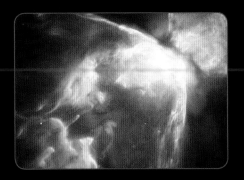

Catalogued as NGC 6302 but better known as the Butterfly or Bug Nebula, this beautiful cosmic cloud looks delicate and fragile but is the scene of incredibly violent activity.

THE BUTTERFLY NEBULA

A dying star has swollen up to many times its original size during its main sequence and is now raging with surface temperatures around 400,000 degrees Fahrenheit. Intense activity has seen it cast off vast shrouds of gas and dust, forming a planetary nebula. Some of the ejected material is racing away at speeds of more than 600,000 mph, crashing into slower moving material ejected earlier from the star. These produce the white regions in the image, taken by the Hubble Space Telescope's Wield Field Camera 3, where the debris collides and produces shock waves.

The Butterfly Nebula is found within the Milky Way galaxy in the constellation Scorpius and lies roughly 3,800 light-years away from the Solar System. The dying star, one of the hottest in our neighbourhood, is obscured by a donut-shaped ring of dense dust called a torus. This constricts the outflow of matter from the star leading to the bipolar, twin lobed shape reminiscent of an hourglass which extends more than two light years across – or almost halfway between the Sun and its nearest neighbouring star, Proxima Centauri. Temperatures of the gas as it travels out into space are estimated at 36,000 degrees Fahrenheit.

ABOVE: This 2004 NASA Hubble Wide Field Planetary Camera 2 image shows the giant walls of compressed gas, laced with trailing strands, of part of the Butterfly Nebula.

THIS PAGE: The Butterfly Nebula's two gigantic 'wings' of bubbling hot interstellar gas form a spectacular sight in this image produced by the Spitzer Space Telescope's Infrared Array Camera (IRAC)

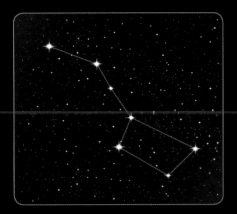

The third largest constellation in the sky, Ursa Major extends over an area of 1279.7 degrees equivalent to 3.1 per cent of the total sky.

URSA MAJOR

THE GREAT BEAR

It borders eight other constellations and is represented as the Great Bear in a remarkably broad and varied range of ancient cultures from the ancient Romans, Finns and Burmese to the Iroquois and Wampanoag native American peoples.

Ursa Major is home to a number of well-observed galaxies including the spiral galaxies, Messier 81, Messier 108 and Messier 109 as well as the Pinwheel Galaxy. I Zwicky 18 is also present. This dwarf irregular galaxy is one of, if not, the youngest known galaxies in existence,

possessing a life time of around four million years, less than a thousandth of the Milky Way's age. In addition, the constellation also contains the Owl Nebula, an 8000 year old planetary nebula some 1,630 light years from Earth. Eight of Ursa Major's stars lie within 32.62 light years of Earth whilst a further 21 have so far been found to form solar systems with exoplanets orbiting around them.

The constellation is home of one of the most famous asterisms (pattern of stars), the Big Dipper (also known as the

Plough). This consists of seven bright stars including Alioth – the brightest star in the entire constellation with an apparent magnitude of 1.77 and a diameter a little over four times that of the Sun. Navigators can make use of the Big Dipper by tracing a line running between two of its stars – Merak and Dubhe – upwards to find Polaris, the pole star, which is a useful indicator of true north.

LEFT: Ursa Major constellation as portrayed in the Zodiacal universe.

Residing in the constellation Ursa Major, along with a host of more famous galaxies and objects, the irregular dwarf galaxy UGC 4459 is shown in this Hubble image. It lies approximately 11 million light years from Earth and contains several billion stars.

This conspicuous and well-known constellation is visible from both the northern and southern hemispheres.

ORION CONSTELLATION

THE HUNTER

It is named after the hunter in Greek mythology and represents a figure of a man with one foot raised and a shield in one hand and a sword or club in the other. Before the Greeks, the ancient Babylonians named it, "the Heavenly Shepherd," and the ancient Sumerians associated the constellation's most prominent stars with their hero, Gilgamesh.

Contained within the constellation are Rigel (Beta Orionis) and Betelgeuse (Alpha Orionis), the seventh and tenth brightest stars in the night sky respectively. Betelgeuse is a young,

(under 10 million years old) red supergiant with a radius at least 900 times greater than the Sun. It is thought that if it were placed in the Solar System, it would extend a considerable distance past the orbit of Mars. Rigel is a blue-white star around 70 times larger than the Sun and with a surface temperature of around 21,000 degrees Fahrenheit. According to the European Southern Observatory, it shines 50,000 times more brightly than the Sun.

The three stars in the middle of the constellation, Alnitak, Alnilam and

Mintaka, form a small asterism, Orion's Belt, running around the waist of the hunter. Just south of the belt is one of the nearest rich stellar nurseries to Earth. The Orion Nebula (also known as Messier 42) is 1,344 light years away and home to hundreds of new stars and protostars as well as more than 150 protoplanetary discs of matter located around infant stars. The nebula is part of the larger Orion Molecular Cloud Complex in which is found the striking Horsehead Nebula.

The Horsehead Nebula peaks out of part of the Orion molecular cloud complex in the constellation Orion. To the left of the Horsehead is the bright star Alnitak (Zeta Orionis). The constellation is best viewed between January and March.

The bull has long been associated with this region of the night sky.

TAURUS CONSTELLATION

THE BULL

In Greek mythology, for example, Zeus transformed himself into a white bull to abduct and carry away a Phoenician princess, Europa, to the island of Crete. The 625 million year old star cluster within the constellation called the Hyades features hundreds of similar stars with the brightest closely-packed within an 8.8 light year wide radius. The five brightest members of this group are past their main sequence and swelling into giant stars. They form the asterism of the bull's head. North-east of these, the stars Zeta Tauri and Beta Tauri (also known as Elnath) mark the tips of the bull's horns. Elnath shines with 700 times the Sun's luminosity. It is considered a chemically peculiar star as it boasts high levels of manganese, silicon and chromium.

Aldebaran is the brightest star in this constellation and one of the top 15 brightest stars overall. Around 65 light years away from Earth, it has swollen to a red giant 44 times the size of the Sun which it shines 500 times more brightly than. It formd the red eye of the bull in the Taurus asterism. The planetary probe Pioneer 10 is travelling in the direction of Aldebaran and is expected to make its closest pass two million years from now. The constellation is home to another scenic and much studied open cluster of stars, the Pleiades and also contains the famous supernova remnant, the Crab Nebula and NGC 1514, a planetary nebula discovered by William Herschel.

The core of the Crab Nebula supernova remnant located in Taurus reveals a turbulent inner region of matter where waves of charged particles and powerful magnetic fields act on the surrounding dust and gas.

Formerly part of an ancient Greek constellation (along with Puppis and Vela) called Argo Navis after the ship sailed by Jason and the Argonauts, Carina, meaning keel of a ship, is found in the southern sky.

CARINA CONSTELLATION

It is home to a number of dramatic and fascinating star clusters amongst them, the Wishing Well Cluster (NGC 3532) which was the very first observational target of the Hubble Space Telescope, the Diamond Cluster and the Theta Carinae Cluster. The latter is also known as IC 2602 or the Southern Pleiades and contains more than 70 stars grouped approximately 540 to 550 light years away. Carina's brightest star is Canopus. Located 313 light years from us,

spectroscopically, Canopus is an F0 type star, hotter than the Sun with a surface temperature of approximately 13,600 degrees Fahrenheit. It shines 14,000-15,000 times more brilliantly than the Sun and with an apparent magnitude of -0.74, is the second brightest star in the night sky.

Home to the unstable star Eta Carinae, Carina is also the location of several truly spectacular nebulae. These include the Carina Nebula (NGC 3372) and NGC 3603,

also known as the Giant Nebula. The latter is the most massive visible cloud of glowing gas and plasma, known as an H II region, in the Milky Way. Described by NASA as a "stellar jewel box", it contains massive young star clusters, amongst the biggest in the Milky Way and is located around 20,000 light years away. It was discovered by Sir John Herschel, son of William Herschel, in 1834.

Looking more like a dazzling firework display, this nebula's central cluster of similarly-aged, huge, hot stars, dubbed NGC 3603, have blown away an enormous cavity in the nebula's shroud of gas and dust. This helps provide this unencumbered view of the cluster which lies 20,000 light years away in the constellation Carina.

This southern hemisphere constellation's name stems from the Portuguese for the dolphinfish and was coined in the 16th Century.

DORADO CONSTELLATION

Whilst the Large Magellanic Cloud is the constellation's most prominent feature, there are others including the highly scenic NGC 1672 spiral barred galaxy. An example of a Seyfert galaxy with an active nucleus, NGC 1672 spans approximately 75,000 light years and features intense star formation particularly of hot blue stars at both ends of its barred region.

Two rapidly moving objects in this constellation excite astronomers, the closest major supernova to us (SN 1987A) which has sent gas hurtling through space at around 10 per cent the speed of light. The constellation also contains one of the fastest-moving stars in the immediate universe. HE 0437-5439 is a massive hypervelocity star (HVS) measured as receding from us at a rate of 449 miles per second or 1.6 million mph. At this velocity, it is not bound by the gravity of a galaxy and is escaping into intergalactic space.

The constellation's brightest star, Alpha Doradus is three and a half times the size of the Sun and has an apparent magnitude of 3.27. It is actually part of a binary system with a smaller companion which is a Cepheid variable. The second brightest star in Dorado as viewed from Earth's viewpoint is R. Doradus with a luminosity approximately 6,500 times that of the Sun. This red giant was measured by a team using the European Southern Observatory's NTT instrument at approximately 370 times the size of the Sun. R. Doradus is eclipsed by RMC 136a1 viewable in the Tarantula Nebula within Dorado whose luminosity is defined in terms of millions of Suns (as much as 8.71 million Suns). This Wolf Rayet star is just two million years old yet is already halfway through its short stellar life and has already cast off around one fifth of its mass.

Located within the Dorado constellation, the Tarantula Nebula features some 2,400 massive stars at its center producing powerful stellar winds and radiation. The Tarantula is a well-known H II region – a region of interstellar atomic hydrogen that has been partly ionized and in which star formation has recently taken place.

CASSIOPEIA CONSTELLATION

Its five most conspicuous stars from our viewpoint form a W-shaped asterism in the sky. Its brightest star, Alpha Cassiopeiae, has an apparent magnitude of 2.24 and was known by its Arabic name of Schedar meaning "breast". The constellation is home to two notable supernovae remnants, SN 1572 or Tycho's Supernova and Cassiopeia A as well as a slowly dissipating 'ghost' nebula, IC 63, and the irregular galaxy, IC 10, the only known starburst galaxy within the Local Group of galaxies that include the Milky Way. A number of star clusters are contained within the constellation including Messier 103 and NGC 7789, the latter also known as Caroline's Rose for its discovery by Caroline Herschel in 1783. Another star cluster, Messier 52, is six light years wide and contains at least 150 members.

Close to Messier 52 can be found the Bubble Nebula (also known as NGC 7635 and Caldwell 11). A hot star there with 10 to 20 times the mass of the Sun is pumping out an intensely powerful stellar wind which has sculpted the surrounding gas and dust into a curved shell or bubble, hence its name. The shell of gas measures 10 light years in width, more than twice the distance from the Sun to its nearest star. The constellation's largest star is PZ Cassiopeiae – a red supergiant with an estimated diameter in excess of 1,100 times the Sun's radius. It is thought to shine with between 213,000 and 270,000 times the intensity of the Sun.

The extraordinary Bubble Nebula as taken by the Hubble's Wide Field Camera 3, was selected as the telescope's 26th birthday image. The seven light year-wide nebula lies 7,100 light years from us in Cassiopeia and is formed from the casting off of material by a star 45 times more massive than the Sun.

A galaxy is a massive, gravitationally bound system that consists of stars, planets, moons, stellar objects including brown dwarfs, nebulae clouds, an interstellar medium of gas and dust and black holes as well as an unknown component described as dark matter.

GALAXIES

These "island universes" in the words of philosopher, Immanuel Kant, vary in size, mostly from 3,000 light years in diameter to over 250,000 light years wide.

Various classification systems exist including the Hubble sequence devised by American astronomer, Edwin Hubble in 1926, colloquially known as the 'tuning fork' for the diagram's two branched form. This classification effectively divides galaxies into three types based on their morphology: spirals, ellipticals and lenticulars. These three categories are further subdivided on how they appear to our observatons. An E0 elliptical galaxy, for example, is almost round whilst an E7 is a long, drawn-out oval shape,

almost like a cigar. A fourth category of galaxy is home to those which have no obvious shape or structure, often due to interactions with other galaxies. These irregular galaxies include our galactic neighbour, the Large Magellanic Cloud.

Estimates of how many galaxies exist continue to be revised and refined, mostly upwards, from thousands to millions in the past, to around 170 billion in the 21st Century. Research derived from the Hubble Space Telescope's data and published in the Astrophysical Journal in 2016, indicate that there could be as many as two trillion individual galaxies – a mind-numbing prospect.

RIGHT PAGE - (TOP LEFT): NGC 6503, a dwarf spiral galaxy with a diameter of around 30,000 light years. (TOP RIGHT): The M101 or Pinwheel Galaxy – a classic spiral galaxy. (CENTER LEFT): NGC 6872, a giant barred spiral galaxy possibly five times the size of the Milky Way. (CENTER RIGHT): NGC 4622, an unbarred spiral galaxy in the constellation, Centaurus. (BOTTOM LEFT): UGC 1322, a giant, isolated galaxy many times the size of the Milky Way. (BOTTOM RIGHT): The NGC 4565 galaxy viewed edge-on by the GALEX mission.

When viewing deep space, telescopes take on the role of time travellers.

HUBBLE
ULTRA DEEP FIELD

Their distant observations of stars, galaxies and other distant objects take astronomers back thousands, millions or billions of years to view these celestial bodies as they were when light first left them and began its journey across space to our viewpoint. The Hubble Space Telescope has periodically taken deep sky field observations, but starting in 2003 went deeper than before. Over two sessions (September 23 to October 28 and December 4 to January 15 the following year) the Hubble's Advanced Camera for Surveys took 800 exposures during 400 orbits of Earth to produce the deepest visible light image of part of the Universe ever attempted by telescope.

Named the Hubble Ultra Deep Field (HUDF), the resulting image stunned professional astronomers and the general public alike. Like a geological core sample, it represented an extremely narrow (just one-seventeeth of a degree across) but extremely deep view of part of the Universe slicing through billions of light years of space. The HUDF was targeted on a narrow portion of the southern hemisphere constellation, Fornax, where the sky is largely empty of stars from within the Milky Way which might mask more distant bodies. A staggering 10,000 galaxies were pictured in the observations, with the larger, more clearly defined elliptical and spiral galaxies one billion light years away. The smallest, reddest galaxies in the HUDF represent some of the most distant known, estimated to be depicted in the image as they were 13 billion years ago, when the Universe was just one-twentieth of its present age.

RIGHT: The galactic zoo that is the Universe is revealed in this portion of the astonishing Ultra Deep Field 2014 image – a composite of separate images taken from 2003 to 2012 using the Hubble's Advanced Camera for Surveys and its Wide Field Camera 3 including ultraviolet data. Some of the dimmest galaxies in shot are over 12 billion years old.

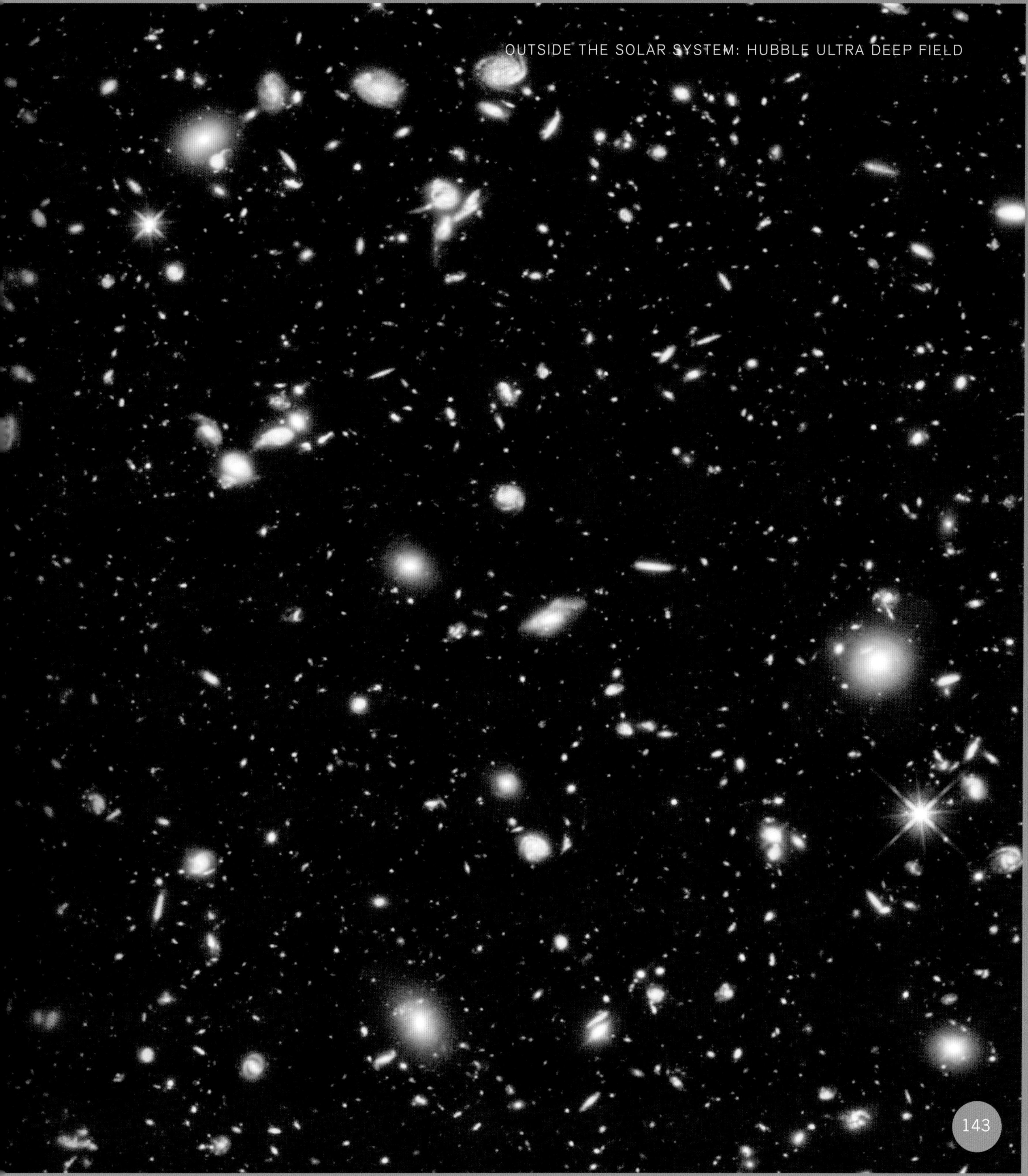

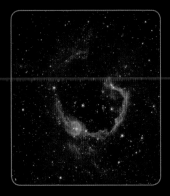

Our galactic home is named after the ancients who likened the spread of white and off-white stars across the night sky to spilt milk.

MILKY WAY

Its name stems directly from the Ancient Greeks and galaxías kýklos, meaning "milky circle" adopted by the Romans as via lactea – the road or way of milk. Our viewpoint located within the galaxy makes it difficult to ascertain its precise shape, size and structure, but astronomers believe it is a barred spiral galaxy. Estimates of its diameter range from 100,000 light years to 200,000 light years, the latter following a 2018 study by astronomers at the Astrofísica de Canarias in the Canary Islands and the National Astronomical Observatories of Beijing, China. Crossing the entire Milky Way at the speed of a current jet airliner (circa 600mph) would take between 111.7 and 223.4 billion years.

The galaxy is a relatively flat disc, warped by its interactions with neighboring galaxies such as the Large and Small Magellanic Clouds. It has a thickness of about 1,000 light years, save for a central bulge about 27,000 light years in width and around six times the average thickness. The galaxy's structure consists of two large, major spiral arms, the Perseus and Scutum–Centaurus arms and a number of minor arms including the Carina-Sagittarius arm rich in star-forming nebula and star clusters. The Solar System is found near another minor spiral arm called the Orion Spur, lying around 26,000 light years from the Milky Way's centre. Our star system orbits about the galactic core at a speed estimated

at 143 miles per second (equivalent to 514,000 mph). Thus, its galactic year or time to complete one orbit is estimated at between 225 and 240 million years.

ABOVE: The Spitzer Space Telescope spies the RCW 79 emission nebula in the southern Milky Way. The 70 light year wide nebula is dated between 2 and 2.5 million years old and lies approximately 17,200 light-years from Earth.

A visible portion of the Milky Way offers a spectacular light show streaming across the night sky. Once can see the faint resemblance to spilt milk amongst its clutches of white stars that gave the galaxy its name.

ABOVE: Dust obscures the centre of the Milky Way and its Central Molecular Zone (CMZ) in this image from the Spitzer Space Telescope. At the heart of the galaxy's core lies a supermassive black hole, dubbed Sagittarius A.

The Milky Way is home to at least 200 billion stars, many located in its arms or central bulge but with significant numbers of older stars found in globular clusters in the halo that surrounds the galaxy. An open cluster close to the galaxy's center, the Arches Cluster, is the most densely packed star region within the Milky Way with as many as 150 stars within a radius of one light year. The Arches is young at less than three million years old but some of the Milky Way's stars extend back relatively close to the theroized dawn of the Universe. HD 140283, or the Methuselah Star is believed to date back at least 13.4 billion years.

More than 50 stars are found within 15 light years of Earth including Barnard's Star, Tau Ceti, Alpha Centauri and the nearest star to us, Proxima Centauri, 4.24 light years away. Even reaching the closest star poses problems beyond technological scope at present; a spacecraft travelling at 35,000 mph would take approximately 81,000 years to arrive at Proxima Centrauri. Whilst the most common stellar object in the Milky Way are thought to be red dwarf stars with masses of 0.075 to 0.5 that of the Sun, the galaxy is home to a number of truly massive stars including VY Canis Majoris and UY Scuti both over 1,000 times the Sun's diameter. According to

NASA calculations, a hypothetical object travelling around the Sun at the speed of light would take 14.5 seconds to orbit its circumference but over six hours to perform the same journey around VY Canis Majoris.

At the centre of the galaxy is a supermassive black hole, named Sagittarius A* and thought to have a mass of approximately four million Suns. Estimates of the entire galaxy's mass are constantly under revision with a 2016 study placing it at 700 billion times the Sun's mass which was uprated in 2019 to around 1.5 trillion solar masses.

This dazzling infrared image from NASA's Spitzer Space Telescope shows hundreds of thousands of stars crowded into the swirling core of our spiral Milky Way galaxy.

Part of the Local Group of galaxies, along with the Triangulum and Andromeda which form the Milky Way's immediate galactic neighbourhood, the Large Magellanic Cloud (LMC) is the largest of the satellite galaxies that are gravitationally bound to our home galaxy.

LARGE MAGELLANIC CLOUD

At around 14,000 light years in diameter, the Herschel Space Observatory estimates it has around one-hundredth of the mass of the Milky Way and may be home to between 15 and 30 billion stars. At approximately 163,000 light years away, it was thought that the LMC was the nearest galaxy to us until the discovery in 1994 of the Sagittarius Dwarf Elliptical Galaxy, which is between 67,000 and 80,000 light years distant.

Located in the night sky in the constellation Dorado close to its borders with Mensa, the LMC is best viewed in the southern hemisphere. Save for the observations of the gifted Arabic astronomer, 'Abd al-Rahman al-Sufi who identified it in the 10th Century CE and named it Al Bakr (meaning the White Ox), the galaxy remained anonymous to northern hemisphere observers until the 16th Century. Two European sailors and explorers, Amerigo Vespucci in 1503-04 and Ferdinand Magellan in 1519 both observed the galaxy on their voyages, the latter lending his name to this galaxy and its companion, the Small Magellanic Cloud.

Although described as an irregular galaxy, observations reveal a partly-formed spiral arm and a significant bar or bulge across the LMC's centre, supporting claims that it was once a barred spiral galaxy distorted by its interactions with other galaxies including the Milky Way. It remains, in the words of Chicago-born astronomer, Robert Burnham Jr, "an astronomical treasure-house" for its rich collection of nebulae including the Tarantula Nebula (also known as 30 Doradus) which at 700 light years across is one of the largest and richest star-forming regions yet observed.

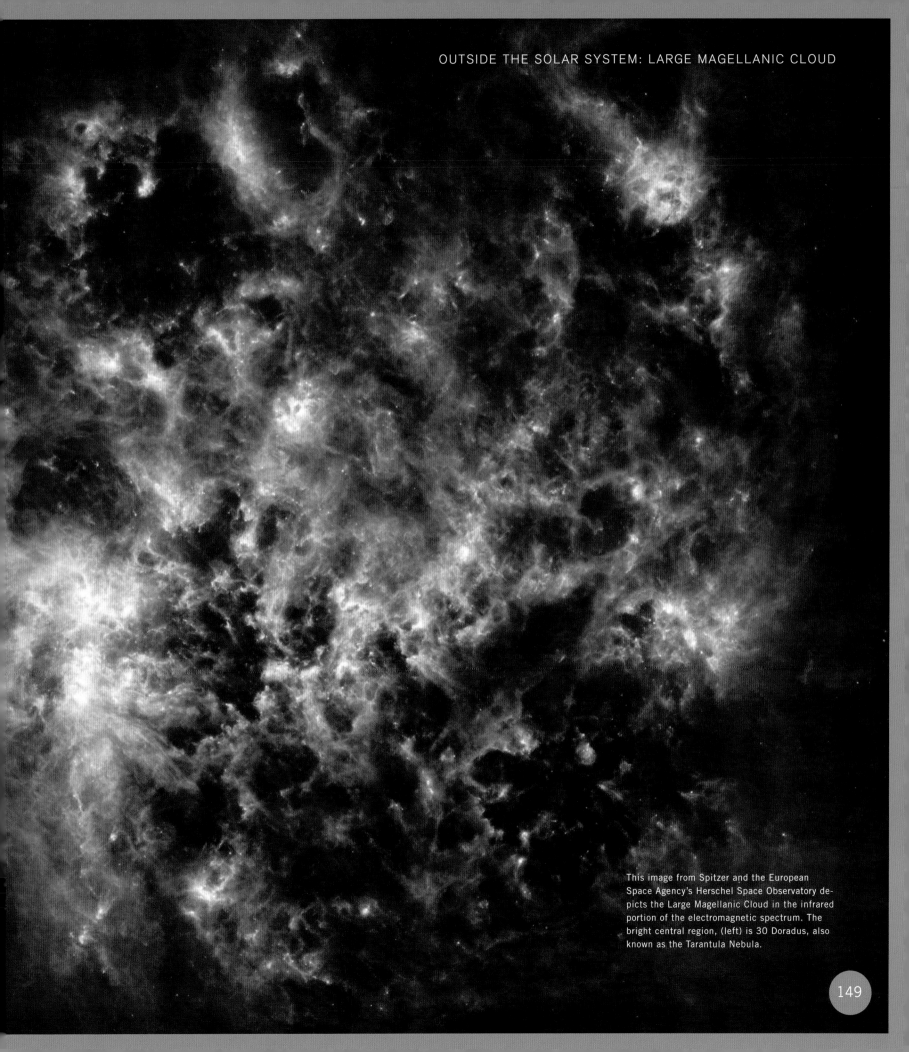

This image from Spitzer and the European Space Agency's Herschel Space Observatory depicts the Large Magellanic Cloud in the infrared portion of the electromagnetic spectrum. The bright central region, (left) is 30 Doradus, also known as the Tarantula Nebula.

The largest galaxy in the Local Group, Andromeda (also known as Messier 31 and NGC 224) is a barred spiral around 220,000-260,000 light years in diameter.

ANDROMEDA GALAXY

AMAZING ANDROMEDA IN RED

Viewed in the infrared (24 microns) part of the electromagnetic spectrum, this image of Andromeda was taken by the Spitzer Space Telescope in 2005. It depicts in magnificent detail, the giant galaxy's broad dust lanes and central bulge containing a dense and compact star cluster.

Angled from our viewpoint in the constellation of the same name and bright at an apparent magnitude of 3.1, according to NASA, it is visible with the naked eye on many nights but provides sumptuous viewing for those equipped with powerful telescopes. In 2015, the Hubble produced its largest single image of the galaxy yet, building up an incredibly detailed picture of a portion of Andromeda as a mosaic of some 7,398 different exposures. The image

was estimated to contain a staggering 100 million stars with the entire galaxy home to more than a trillion. Some are contained in the 460 globular clusters identified within the galaxy.

A colourful, cannibalistic past saw Andromeda's powerful gravity draw in and consume a number of former galaxies. It continues to exert a relentless bullying force on many of the 14 dwarf and satellite galaxies within close vicinity

including Messier 110 and Messier 32. Andromeda lies just over 2.5 million light years from the Milky Way...for the moment. The two galaxies are moving towards each other at a blueshift rate estimated at 68 miles per second. A head-on collision is predicted, and in 2019, was estimated by the European Space Agency's Gaia space observatory as being an event 4.5 billion years away.

THIS PAGE: Hot stars burn brightly in this ultraviolet image of the Andromeda Galaxy taken by NASA's Galaxy Evolution Explorer (GALEX) space telescope. Dark blue-grey lanes of cooler dust contrast with the bands of blue-white containing large numbers of hot infant stars whilst contained in the central region are a collection of cooler, older stars.

The Cigar Galaxy was discovered, along with its neighbor Messier 81 or Bode's Galaxy, by German astronomer Johann Elert Bode in 1774.

M82
CIGAR GALAXY

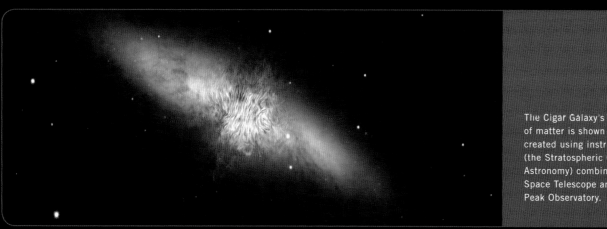

The Cigar Galaxy's powerful outflowing of matter is shown in red in this image created using instruments aboard SOFIA (the Stratospheric Observatory for Infrared Astronomy) combined with the Spitzer Space Telescope and, on Earth, the Kitt Peak Observatory.

This "nebulous patch" according to Bode was catalogued by Charles Messier as Messier 82 and studied heavily since. Turbulent gravitational interactions between the two galaxies have helped trigger unusually high levels of star formation, making the Cigar galaxy a classic example of a starburst galaxy. Whilst at 37,000 light years across it is under a third of the size of the Milky Way, its rate of star formation is far higher; a study of over 190 massive star clusters in a central portion of the Cigar Galaxy revealed stars forming at a rate 10 times faster than our entire galaxy. Radiation and energetic particles from new stars shape and compress surrounding gas leading to further star formation.

Located between 11 and 12 million light years from Earth in the constellation Ursa Major, the galaxy has an apparent magnitude of 8.4. An extremely powerful stellar wind emanates from the galaxy and was measured by NASA scientists in 2018 using the SOFIA airborne infrared telescope. They estimated it contained the equivalent of between 50 and 60 million solar masses of gas and dust. The galaxy has yielded further surprises. Four years earlier, whilst investigating a powerful supernova, SN 2014J, in the galaxy, astronomers discovered M82X-2, the brightest pulsar yet identified. NASA's Nuclear Spectroscopic Telescope Array (NuSTAR) measured its X-ray output as the energy equivalent of ten million Suns.

A false colour view of the mysterious Cigar
Galaxy depicts visible light data in green and
orange, X-rays in blue and infrared emissions,
most notably from the top of the galaxy, in red.

This majestic spiral galaxy's orientation
almost entirely flat on to our viewpoint
makes it a galactic visual feast.

PINWHEEL GALAXY

The galaxy was discovered by Pierre
Méchain, a colleague and French
compatriot of Charles Messier's in 1781
in the constellation Ursa Major and
given the Messier catalog designation
M101. Spanning a portion of space
estimated at 170,000 light years across,
the galaxy is thought to be home to over
a trillion stars, 100 billion of which are
ascertained to be Sun-like in character.
It was thought to be between 25 and 27
million light years distant but recent

revisions have brought that figure down
to 21 million.

The galaxy's clearly defined spiral arms
are sprinkled with large regions of
nebulae in which new star formation is
occurring. More than 3,000 clusters of
hot, blue newly formed stars have been
observed and help illuminate the galaxy's
arcing arms. Although categorised
as a 'grand design' spiral galaxy (a
galaxy with a clearly defined and

organised spiral structure), the Pinwheel
is strikingly asymmetrical which
astronomers believe is to do with the
interactions between the Pinwheel Galaxy
and its five companion galaxies Holmberg
IV NGC 5204, NGC 5474, NGC 5477 and
NGC 5585, the latter also a spiral galaxy.
It is best observed in the spring months
in the northern hemisphere,
especially April.

LEFT: Young and old stars are relatively evenly distributed amongst the tightly-wound spiral arms of the M101 galaxy in this archive image.

THIS PAGE: Ten years' worth of Hubble observations as well as ground-based observations have gone into this detailed view of M101. The galaxy's relatively small central bulge has a mass of around three billion Suns and has very little starbirth occurring, in great contrast to its spiral arms.

Lying over 28 million light years from Earth at the southern edge of Virgo cluster, this spiral galaxy was given its common name (it is also known as Messier 104 and MGC 4594) for its slight resemblance to a broad-brimmed Mexican hat.

SOMBRERO GALAXY

The Sombrero Galaxy when viewed in infrared as well as visible light in this image looks less like a Mexican hat and more like a bull's eye. Observations by the Spitzer Space Telescope show how the galaxy's disc is warped slightly, a possible result of a previous encounter with another galaxy and its immense gravitational pull. Located in the constellation, Virgo, the Sombrero is thought to harbour a truly gigantic supermassive black hole at its centre believed to be in the order of one or more billion times the mass of the Sun.

Its unusual look is purely about viewpoint. From our perspective, the galaxy is viewed a mere six degrees away from perfectly edge-on, revealing in great detail the broad dust lane that encircles the galaxy and the bright, bulbous core partially illuminated by the galaxy's 2,000 or so identified globular clusters in its 50,000 light year diameter. The galaxy's magnitude of +8 makes it relatively bright but its distance means it is just out of view of the naked eye.

The galaxy was first spotted by French astronomer, Pierre Méchain in 1781 but was somewhat lost to history, as it was noted by star cataloguer Charles Messier but not included in his official list of Messier objects until 1921. It was used, however, in 1912, to measure redshifts (bodies moving away) by Lowell Observatory's Vesto Melvin Slipher in 1912 and was one of, if not the first, galaxy to be found to exhibit a significant recession velocity. Slipher estimated that the Sombrero was moving away at a velocity of over 680 miles per second, equivalent to 2.4 million mph, lending weight to the theory later advanced by Edwin Hubble concerning the expansion of the Universe.

The ethereal ghostly glow emanating from the Sombrero Galaxy's bulbous core is contrasted by the dark, thick dust lane encircling the galaxy in this image comprised of six separate images merged together by the Hubble Space Telescope's Heritage Team.

Also known by the number 4258 in the New General Catalogue of Nebulae and Clusters of Stars, Messier 106 was discovered by Pierre Méchain in 1781.

MESSIER 106

It is found in the constellation, Canes Venatici and lies around 22 to 23 million light years away, making it one of the Milky Way's closest spiral companions. At around 135,000 light years in diameter, it is comparable in size to our home galaxy. The galaxy has a relatively bright apparent magnitude of 9.1 and, according to NASA, can be spotted with a small telescope or even high-powered binoculars.

At the start of the 1940s, Harvard-educated astronomer Carl Seyfert began work on studying particularly bright galaxies whilst working at the Mount Wilson Observatory. He discovered that some spiral galaxies had far brighter and energetic cores than typical and

were populated by hot glowing clouds of hydrogen, helium and other elements. Today, we know that powering such active galaxies, now called Seyfert galaxies, are supermassive black holes at their center. As gas is drawn and spirals towards the black hole, it heats up and emits powerful and destructive radiation. Messier 106 is one such Seyfert galaxy.

Most spiral galaxies have two major spiral arms. Messier 106, though, is different. Two of its four spiral arms are full of stars but an extra set of arms curve out and are packed full of hot gas. As the official Hubble Space Telescope website states, "The extra arms appear to be an indirect result of jets of material produced by the violent churning of matter around

the galaxy's core." Partly as a result, Messier 106 is more visible in radio, X-ray and microwave extents than in the visible light portion of the electromagnetic spectrum. Many of the more detailed images of this galaxy involve merging data from different observatories into a single composite image.

ABOVE: An incredible view of the M106 galaxy combines Hubble imagery with observations made by ground-based telescopes based in New Mexico. Its glowing central core can be clearly seen as can the anomalous arms of gas (shown in red).

THIS PAGE: Radio wave analysis from the Karl Jansky Very Large Array along with Hubble, Chandra and Spitzer observations have been combined to produce this detailed image of M106. Its anomalous arms (shown in purple) comprise hot gas rather than large numbers of stars and are the indirect products of violent activity close to the galaxy's supermassive black hole.

The small, southern hemisphere constellation, Corvus (meaning Raven) is known for its four brightest stars forming a quadrilateral that can be easily observed in the night sky.

ANTENNAE GALAXY

Not so easy to spot and requiring significant resolving power are a pair of galaxies, the barred spiral NGC 4038 and the spiral galaxy NGC 4039 locked in seemingly slow-motion collision more than 45 million light years away.

First discovered by William Herschel in 1785, the galaxies are so nicknamed due to their long streamers of material which bear a fleeting resemblance to elongated insect antennae. These are the result of the collision process which is thought to have begun more than 100 million years earlier. As the galaxies, with a collective diameter of 500,000 light years, crunch into each other, dust, gas and stars are ripped from their location and ejected as streamers. The crash has also compressed and heated up dust and gas inside the galaxies, prompting furious and violent periods of star formation, much of which is occurring in tight clusters of activity. Many of the newest stars are classified as huge but short-lived blue hypergiants with a lifespan of just 10 million years.

In total, the two galaxies are thought to contain around 300 billion stars. The cores of the two galaxies are an estimated 30,000 light years apart but this distance will gradually shrink as the two galaxies completely merge, a process that astronomers believe will occur within the next 300 to 500 million years. The end result is likely to be a large or giant elliptical galaxy.

NGC 4038 (top) and NGC 4039 (bottom), collectively known as the Antennae galaxies, are viewed here in all their violently colliding glory, via observations taken between 1999 and 2005.

This picturesque galaxy, a favourite with amateur astronomers, is known by many names including Messier 51A, NGC 5194 and Rosse's Galaxy after William Parsons, the third Earl of Rosse.

WHIRLPOOL GALAXY

A composite image of the Whirlpool Galaxy, showing the majesty of its spiral structure involved over 250 hours of observations by the Chandra X-ray Observatory depicting hot X-ray emissions in purple with Hubble data providing other colours.

Formerly known only as a fuzzy nebula, Parsons' observations in 1845 revealed the galaxy's spiral construction using the world's largest telescope of the time, the 72 inch aperture Leviathan of Parsonstown which featured a five inch thick mirror and weighed over 12 tons.

Located in the constellation Canes Venatici between 26 and 31 million light years away, the Whirlpool is a classic example of a grand design spiral. Its elegant, long and winding spiral arms curving gracefully from its central core which contains a supermassive black hole. With an apparent magnitude of 8.4, the Whirlpool can be spotted using binoculars. Three supernovae (in 1994, 2005 and 2011) have been spotted within the galaxy – an unusually frequent occurrence.

The Whirlpool is not alone. It travels through space with a companion galaxy, NGC 5195 in close attendance. The smaller, irregular galaxy is passing behind the Whirlpool, a journey hundreds of millions of years in the making. Both galaxies' gravity has exerted influence on the other with NGC 5195 distorted and pulled out of shape but in turn acting on the Whirlpool's arms, causing gas to compress into dense clouds which collapse to form new stars packed into star-forming regions revealed in pink on the image.

The stunning spiral structure of the Whirlpool Galaxy stand out in this Hubble image. The galaxy is approximately 60,000 light years in diameter and can be viewed faintly with regular binoculars or seen in more detail with a good amateur telescope..

The year 1992 marked the discovery by US astronomer Dale Frail and Polish astronomer Aleksander Wolszczan of the first known extrasolar planets or exoplanets, orbiting the pulsar PSR B1257+12.

EXO PLANETS

Three years later came the announcement of Pegasi 51 b, the first exoplanet orbiting around a main sequence star. The pace of discovery has been rapid ever since, so much so, that as of May 2019, there are 3,940 confirmed exoplanets with a further 3,504 candidates cited by NASA as requiring further investigation and verification before confirmation.

Exoplanets can be extraordinarily difficult to observe compared to their parent star; only 1.1 per cent of all discovered exoplanets have been found via direct imaging. Instead, astronomers use micro lensing, radial velocity (where shifts in a star's light color is caused by its wobble due to orbiting exoplanets) or by transit, observing the exoplanet passing directly in front of the star; more than three-quarters of all exoplanets have been discovered using the transit method.

Exoplanets exhibit huge variety in their size, orbits and surface conditions. Some experience sub-zero temperatures whilst Kepler-70b's surface is hotter than the Sun. J1719-1438 b completes its orbit in 2.2 hours whilst 2MASS J2126−8140's orbit lasts more than 900,000 years. At the time of writing, HD 100546 b is the largest known exoplanet with a diameter, estimated by NASA at more than six times that of Jupiter. Many exoplanet hunters, utilising data gathered by devices like the Kepler space telescope and new space observatories including NASA's TESS and the ESA's CHEOPS, are less focused on these superlatives and more on finding planetary systems orbiting stars within habitable zones where conditions might exist that could support life.

The seven planets that orbit the cool red dwarf star,
TRAPPIST-1 are all Earth-sized and terrestrial, rocky bodies.
Trappist 1g (second from right) is the largest of the planets,
approximately 1.15 times the radius of Earth but is less dense,
suggesting that it might harbour water in some form amongst
its rocky surface and mantle.

EXPLORING
SPACE

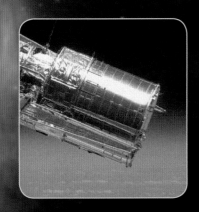 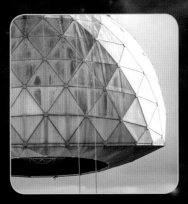  

Arrays are two or more radio telescopes working together to pick up faint radio waves from objects in the Universe.

VERY LARGE ARRAY

They can have the advantage of easier construction than a single giant dish and more flexibility in that they can be mounted and moved to change the viewing target or level of resolution and depth and detail of information they gather in. The Very Large Array (VLA), named in 2012 after a pioneer of radio astronomy, Karl G. Jansky, occupies an isolated location in the New Mexico desert, at an elevation of 6970 feet, away from sources of noise and interference.

After Congress approved funding in 1972, the VLA began construction the following year and went operational in 1980, as part of the National Radio Astronomy Observatory (NRAO). It consists of 27 dish antennae arranged in a Y shaped pattern. Each dish is 82 feet in diameter and weighs 230 tons. When working together, they have the potential to operate with as much information-gathering power as a single dish 22.6 miles in diameter. Each of the VLA's dishes contains eight receivers to process radio waves. The electronics inside each dish are cooled by helium gas to -432 degrees Fahrenheit.

Four times a year typically, the dishes go on the move. They are fitted to altitude-azimuth mountings which one of two 90 ton rail transporter vehicles move along tracks to lengthen or shorten each arm of the array from two-thirds of a mile up to 23 miles. Changing the spacing the dish antennae allows varying observation types to take place. The farther apart the dishes are, for example, the better the telescope's resolution, or ability to distinguish between two close objects. When the dishes are closer together, the resolution is less, but the telescope may obtain more information about characteristics such as the star, galaxy or other object's temperature and brightness.

The Milky Way looms brightly over one of the Very Large Array's antenna dishes in the New Mexico desert.

169

THIS PAGE: A cluster of VLA antennae each on their steel mountings featuring a powered azimuth bearing which can turn the dish at a rate of 40 degrees per minute.

The VLA working in tandem with Hubble's Wide Field Camera 3 produced this arresting image of the elliptical galaxy, Hercules A and its spectacular jets powered by the gravitational energy of a supermassive black hole at its core.

The VLA's productivity has been astounding, with more than 11,000 observing projects run at the facility so far, and its discoveries legion. Amongst the most famous are its discovery of microquasars and in 1991, the locating of water ice on Mercury in permanently shaded craters, despite the planet sometimes experiencing temperatures as high as 800 degrees Fahrenheit. It was also used to communicate with the Voyager 2 spacecraft as it passed Neptune in 1989 as well as capturing radio images of pebble-sized chunks of matter outside the Solar System in the first stages of planetary formation. Data derived from its radio observations are often combined with images from other observatories operating in different areas of the electromagnetic spectrum to build up detailed composite views of a galaxy or other object.

New work and findings kept on coming in the 21st Century including, in 2011, the identification of a supermassive black hole with a mass a million times more than the Sun in the irregular star-forming galaxy, Henize 2-10, some 30 million light years away. The following year, the VLA received a major upgrade in its electronics and processing power before in 2017 it began a far-reaching seven year project called VLASS to make three complete scans of the entire sky. VLASS is expected to produce the sharpest ever radio views of such a large portion of the sky, and may detect as many as 10 million distinct radio-emitting celestial objects, about four times as many as are known currently.

Perched inside the depression formed by a large karst sinkhole in Puerto Rico, the Arecibo Observatory was until 2016, the biggest single dish radio telescope in the world.

ARECIBO OBSERVATORY

Constructed at a cost of $9.3 million over a three year period and opening in 1963, the 1000 foot diameter dish surface was pieced together from 38,788 perforated aluminum 6 feet by 3 feet panels to form a spherical reflector, gathering in radio waves from space and focusing them on the receiver suspended 452 feet above the dish on a series of cables attached to a platform weighing around 900 tons. To get a better idea of scale, the dish covers an area of 18 acres – almost equal to 24 full-sized football fields.

Although now surpassed in size by China's monumental FAST (Five hundred metre Aperture Spherical Telescope), there remains an argument that Arecibo still reigns supreme simply due to the weight of it body of work and the discoveries made using the device. It was initially constructed on the US Air Force's behest primarily to use radar to study the Earth's ionosphere, but very quickly it became apparent that astronomers and scientists had a powerful tool at their command that could reach and

investigate far further. Arecibo has been used to recalibrate our understanding of the length of the planet Mercury's days, provide evidence of the existence of neutron stars, discover binary pulsars and detect Earth-crossing asteroids. In addition, the very first exoplanets were discovered orbiting a pulsar first detected in 1970 using Arecibo.

ABOVE: Arecibo's Gregorian reflector, installed in 1997, weighs approximately 75 tons and is suspended 450 feet above the dish.

Arecibo's giant dish located in the rugged northern Puerto Rican countryside.

The largest, most powerful launch vehicle to ever propel a mission successfully into space, the Saturn V was designed for the Apollo missions.

SATURN V

Developed at NASA's Marshall Space Flight Center in Huntsville, Alabama, but involving contractors from a staggering 20,000 different companies and organisations, the three-stage vehicle towered 363 feet tall, 60 feet taller than the Statue of Liberty and more than three times the height of the Titan II used to launch the Gemini spacecraft. The vehicle's huge first and second stages, built in New Orleans and California, were transported on barges down the Mississippi or through the Panama Canal. Weighing 6.2 million pounds, each Saturn V was assembled in the VAB (Vehicle Assembly Building) at Kennedy Space Center and rolled to its launch pad on the back of a crawler-transporter larger than a baseball diamond.

A Saturn V was made up of three stages, each jettisoned after its fuel was exhausted. The first stage, featured five Rocketdyne F-1 engines. Each consumed 3,945 pounds of liquid oxygen (LOX) and 1,738 pounds of RP-1 fuel per second. Together, the five engines generated 7.6 million pounds (34.5 million newtons) of thrust at launch and burned for 168 seconds by which time, the vehicle was moving at a rate of approximately 7,500 feet per second and at an altitude of 36 nautical miles. The second stage's five J-2 rocket engines took over, firing for around six minutes to reach an altitude of 109 miles and velocity of 15,647 mph before the S-IVB or third stage's single J-2 engine took the craft to translunar injection. Thirteen Saturn V launches, all successful, were made, twelve involving test flights or the Apollo mission with the last in 1973, sending the Skylab space station into orbit.

The five F1 engines mounted in the base of the S-1C first stage of a Saturn V on display at the Kennedy Space Center.

A Saturn V launch vehicle (SA-502) lifts off from Kennedy Space Center in April 1968 carrying the Apollo 6 unmanned test mission. The first manned mission, Apollo 8, would be launched by another Saturn V in December of that year.

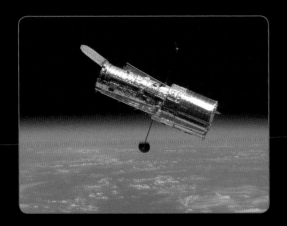

One of NASA's four Great Observatories and named after the pioneering American astronomer Edwin Hubble, the Hubble was launched in 1990 and has transformed our understanding of many aspects of astronomy and space science.

HUBBLE SPACE TELESCOPE

A flaw in the reflecting telescope's primary mirror, leading to fuzzy images, was corrected with the 1993 Space Shuttle STS-61 mission which fitted a giant contact lens, the Corrective Optics Space Telescope Axial Replacement (COSTAR) to the telescope. It was the first of five service missions which have upgraded instruments and added some 3,000 pounds to the Hubble's initial 24,000 pound weight.

Taking 95 minutes to complete an orbit at its current altitude of 340 miles, the 43.5 foot-long telescope features a 7.9 feet diameter primary mirror capable of gathering in 40,000 times more light

than the human eye. The telescope is able to lock onto a target without deviation of more than 0.007 of an arcsecond – equivalent to the width of a human hair at a distance of one mile. Untroubled by light pollution or the distorting effects of Earth's atmosphere, the Hubble is able to make pin-sharp observations and take images of both Solar System and deep space objects.

Powered by two 25 foot long solar panels, the Hubble generates approximately 10 terabytes of new data per year. Its data and imagery has led to the publication of more than 15,000 scientific papers and a vast array of discoveries

from determining the age and rate of expansion of the Universe to investigating star formation and discovering new moons around Pluto. In the words of John Hopkins University astronomer, Mario Livio, "there is essentially no area of astronomy and astrophysics where [the] Hubble didn't contribute something quite significant."

ABOVE: The Hubble Space Telescope orbits 353 miles above Earth's surface in this image taken shortly after its second servicing mission in 1997. The mission upgraded several instruments including replacing a tape recorder with a solid state device to record data from the Hubble's observations.

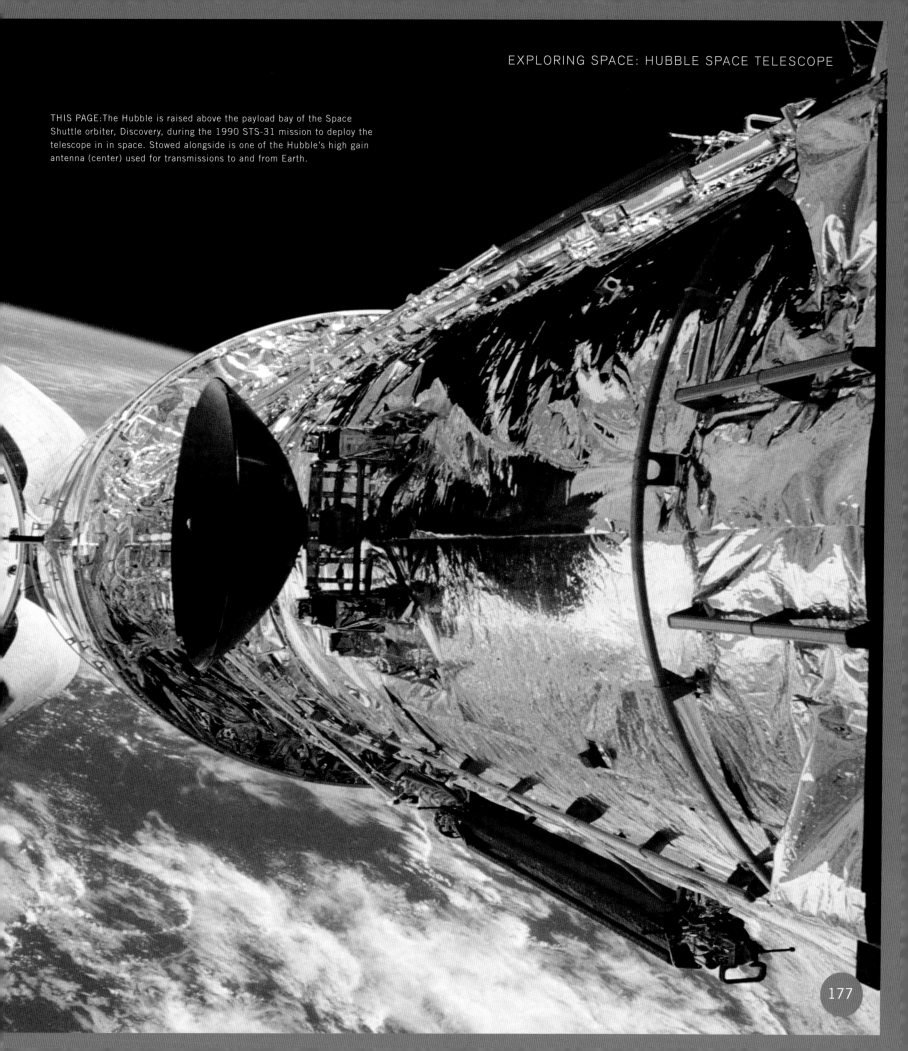

THIS PAGE:The Hubble is raised above the payload bay of the Space Shuttle orbiter, Discovery, during the 1990 STS-31 mission to deploy the telescope in in space. Stowed alongside is one of the Hubble's high gain antenna (center) used for transmissions to and from Earth.

177

Launched in 2003 on an unusual orbit where it trails behind Earth and drifts further away (about one tenth of an AU per year), Spitzer is the fourth and last of NASA's Great Observatories, along with the Hubble, Chandra and the Compton Gamma Ray Observatory.

SPITZER SPACE TELESCOPE

Tasked with observing in the infra-red portion of the electromagnetic spectrum, Spitzer can detect cool stars, exoplanets and dusty stellar nurseries where new stars are forming as well other bodies in space obscured by dust or rendered invisible to other telescopes.

While its field of view changes throughout the year, at any given time the telescope can view around one-third of the sky gathering in infra-red energy using its 33 inch diameter mirror made of beryllium and cooled to -459 degrees Fahrenheit, just above absolute zero, to aid the detection of the faintest infra-red traces from distant objects. Spitzer transfers images and data from its instruments to Earth using a wireless telemetry link with NASA's Deep Space Network using eight gigabits of on board data storage as a buffer.

In 2009, the same year that it discovered an additional ring around Saturn 300 times the planet's diameter, the telescope exhausted its supply of liquid helium leading to the malfunctioning of several of its instruments. The telescope has soldiered on and although not designed as an exoplanet hunter, Spitzer has made significant contributions to the task. In 2009, it produced the first weather map of an exoplanet – HD 189733b – and six years later found one of the most distant identified exoplanets – a gas giant dubbed OGLE-2014-BLG-0124Lb some 13,000 light years away from Earth.

ABOVE: This image from Spitzer shows the star-forming features nicknamed, "yellowballs" found in the W33 region of the Milky Way. Each is larger than the entire Solar System.

ABOVE: Technicians put the final touches to the Spitzer Space Telescope (formerly, the NASA Space Infrared Telescope Facility) at the Lockheed Martin Aeronautics plant in Sunnyvale, California. The telescope was launched on August 25, 2003 and has vastly exceeded its expected operational period of 2.5 to 5 years.

Extravehicular Activity or spacewalks in popular parlance are mini missions within a manned mission where one or more astronauts head outside of their craft and into space protected by a complete life support system in a suit - in the case of NASA astronauts, an Extravehicular Mobility Unit or EMU.

EVAs

This features a multi-layered suit able to repel micrometeoroids and protect against space radiation and a PLSS backpack which circulates oxygen around the suit and supplies electrical power to the EMU's cooling, communications and health monitoring systems. NASA astronauts also don an additional safety system - the Simplified Aid For EVA Rescue (SAFER) which contains 24 small jet thrusters and is steerable, propelling an astronaut to safety in the emergency.

Originally performed as experiments and as preparation for manned missions to the Moon, spacewalks have developed into valuable tools in space exploration

enabling astronauts to conduct and retrieve external experiments, to capture or upgrade satellites and space telescopes or to construct, modify or repair space stations. The first spacewalk occurred on March 18, 1965 when Alexei Leonov left his Voskhod 2 spacecraft for a 12 minute sojourn into space attached to his craft by a 17 foot long tether. Two and a half months after Leonov's mission, Ed White performed the first American spacewalk, leaving Gemini 4 for a total of 23 minutes. The following year, Edwin 'Buzz' Aldrin performed three spacewalks, totalling 5 ½ hours, proving that astronauts could perform genuinely useful work in space. EVA numbers and durations mounted

especially with the construction of the International Space Station. By 2018, 212 EVAs had been conducted from the ISS, the longest of which, performed by James Voss and Susan Helms in 2001 lasted eight hours and 56 minutes.

ABOVE: NASA astronaut Robert L. Curbeam, Jr. prepares to replace a faulty TV camera on the exterior of the International Space Station during a 2006 EVA which lasted 6 hours and 36 minutes. Swedish astronaut, Christer Fuglesang (unpictured) also participated.

STS-114 mission specialist Stephen K. Robinson, performs an EVA anchored to the foot restraint of the International Space Station's Canadarm2 robotic arm in 2005. Robinson became the first astronaut to perform an in-flight external repair to a Space Shuttle during this EVA.

It doesn't get any bigger than this. Dubbed, "humanity's home in space," by NASA, the International Space Station (ISS) is four times larger than any previous space station and five times larger than Skylab.

INTERNATIONAL SPACE STATION

It spans an area equal to a gridiron field and has the same amount of pressurised internal space (32,333 ft3) as a typical five or six-bedroomed house. The space station gained an additional 565 cubic feet with the installation of an inflatable BEAM module trialled at the station since 2016. Stretching 357 feet, 6 inches at its longest point, the ISS has a mass of approximately 925,335 pounds or roughly 320 cars and orbits Earth at a current altitude of 250-254 miles, travelling at 17,100 mph.

The idea of the ISS began back in 1982 with a Space Station Task Force set up at NASA headquarters. A gradually evolving design was bolstered as international partners came on board including Japan, Canada and the European Space Agency in 1985-88 followed later by Russia, Sweden and Switzerland. In 1998, the very first ISS component, the Russian-built Zarya (Sunrise) module travelled into space on board a Proton-K rocket. This 41.2 feet long functional cargo block, contained three docking ports and 16 external fuel tanks capable of holding 5.4 tonnes of propellant fuel. It was followed, less than a month later, by Unity, a linking node carried on board the Space Shuttle's STS-88 mission and in 2000, the Zvezda (Progress) service module.

With Zvezda, came habitation and life support facilities for human crew members and the ISS has been continually inhabited ever since, today most typically by crews of six astronauts. The space station's current great extent was the result of an epic construction program featuring in excess of 120 EVAs and with the large modules and station components delivered on 42 assembly flights, 37 on Space Shuttles and five on Russian Proton/Soyuz launch vehicles.

ABOVE: NASA
Astronaut Michael Fincke, Expedition 18 commander, works inside the Kibo laboratory airlock of the International Space Station in 2009.

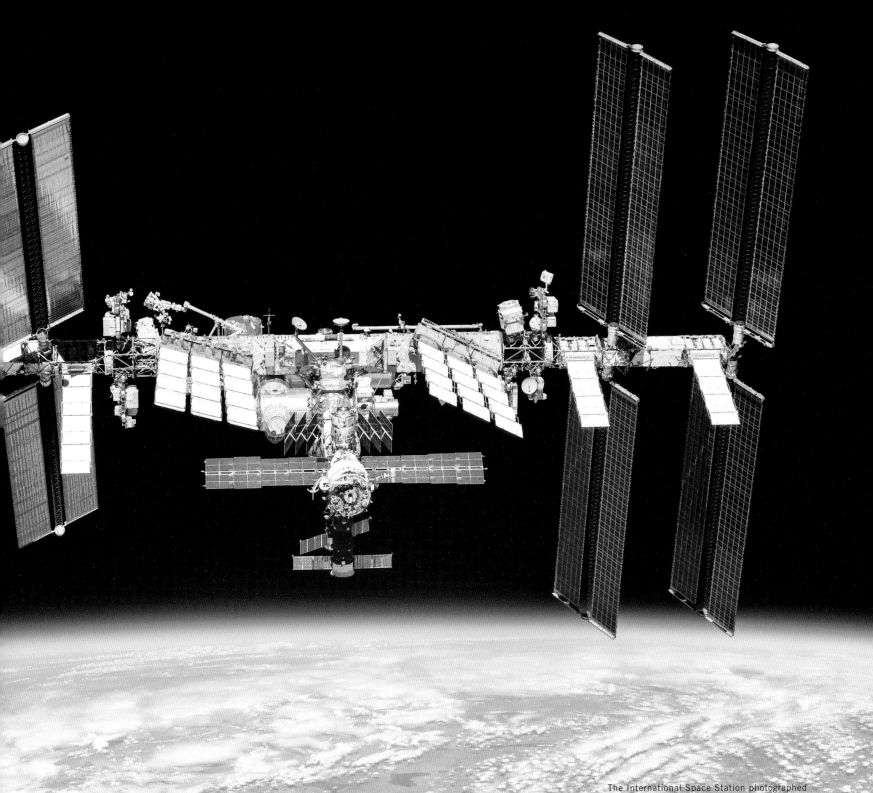

The International Space Station photographed by Expedition 56 crew members from a Soyuz spacecraft after undocking in October 2018.

The International Space Station's advanced modular construction provides unparalleled space and facilities for experiments and other space science investigations. Its integrated truss structure (ITS) forms the space station's spine and features the mobile base system which allows robot arms such as Dextre and the 58 foot-long Canadarm2 to move along the space station's length. Four large pairs of solar array wings (SAWs) measure 240 feet long and each contains 32,800 solar cells. Together, these are rated to produce as much as 110 kilowatts of electricity and are mounted on gimbals so can turn to align with the Sun. When in direct sunlight, only 40% of this capacity is used to run the ISS; the remainder is used to charge electrical equipment batteries. Excess heat in the ISS is vented out into space via large thermal radiators which are part of the Active Thermal Control System.

Attached to the truss or other modules are linking connectors called nodes and an array of different modules. Made of aluminium and weighing 32,000 pounds, the Destiny Laboratory is NASA's main research and science module on the ISS. It features an optical grade window through which high resolution video of Earth and space can be shot. The Kibo module (Japanese for 'hope') is Japanese space agency JAXA's first human-rated space facility. The 36.7 foot-long, 14.4 foot diameter research module can accommodate up to four astronauts at a time and comes complete with its own 32.8 feet-long robotic arm.

William Shepherd, Yuri Gidzenko and Mir space station veteran Sergei Krikalev formed the first expedition to the early ISS, staying for 136 days onboard. Krikalev would return as commander of Expedition 11, becoming the first person to twice visit the ISS. In between, NASA astronaut Susan J. Helms became the first woman to live on the space station. As of mid-2019, 18 different nations have provided ISS inhabitants with the US providing the lion's share, some 149 of the 230 astronauts on board.

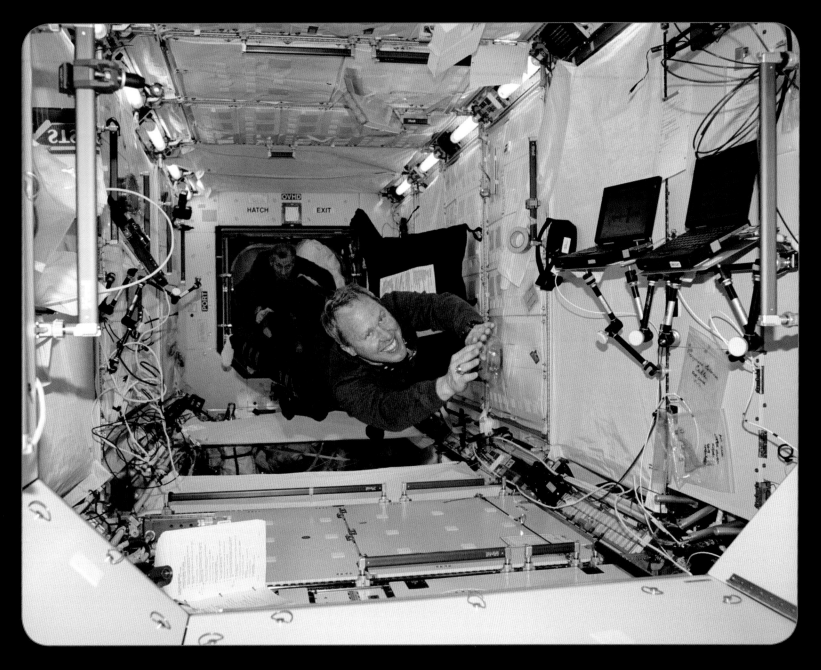

LEFT: This photo taken from the departing Space Shuttle Discovery in March 2001, shows the partially-completed ISS in its early stages before the arrival of various modules including Kibo from Japan and the ESA's Columbus lab.

ABOVE: This STS-98 mission photograph shows astronauts Thomas D. Jones (foreground) and Kenneth D. Cockrell float inside the newly-installed Destiny Laboratory Module aboard the International Space Station in 2001. The module is the cornerstone of American space-based research aboard the orbiting space station.

The biggest robotic rover so far landed on Mars in 2012 following a 352 million mile journey.

CURIOSITY ROVER

The rover - housed in a protective aeroshell - was gently winched down onto the Martian surface on the end of a 25ft cable by an innovative rocket-powered skycrane. Compared to the first rover on Mars, the 11 inch high Sojourner, Curiosity is hefty and the size of a car measuring 9 feet 10 inches long, 9 feet, 1 inch wide and 7 feet high. The rover's 20 inch diameter wheels and bogie suspension system enable it to clamber over obstacles and inclines up to 45 degrees, traversing a maximum of 100 feet per hour. Weighing 2000 pounds, its power requirements are met by a multi-mission radioisotope thermoelectric generator, which produces 2.5kWh of electricity per day from the heat of plutonium-238's radioactive decay; the rover's 11 pounds of plutonium are enough for 14 years of operation.

Curiosity is well equipped with sensors and cameras, 17 of the latter, many used for the robot's autonomous navigation and hazard detection whilst the Mastcam can take high resolution still images and high definition 10 fps video. In its first year on Mars, the rover transmitted more than 36,000 images back to Earth. Amongst its large array of scientific instruments are a complete weather station mounted to the mast (REMS), two radiation detection sensors and four spectrometers to analyse the composition of the rocks and soil the rover studied for signs of life or clues to Mars' past.

One spectrometer, part of the ChemCam package, analysed rock vaporized by the instrument's infrared laser. Wielding its sophisticated array of instruments, Curiosity has identified the presence of sulfur, nitrogen, boron, phosphorus, oxygen, carbon other key chemical ingredients for life on Mars, one of many discoveries for a rover that celebrated its 2300th Sol (Martian day) in action on the Red Planet in 2019.

ABOVE: Curiosity fires the laser in its ChemCam instrument to vaporise a small sample of Martian rock for analysis in this artist's impression.

Curiosity takes a selfie on Mars on May 11, 2016 using its Mars Hand Lens Imager (MAHLI) camera. At this point the rover had been operation for 1,338 sols (Martian days) and was located on a plateau on Mount Sharp taking drilled samples of the surrounding rocks for on board analysis.

Our knowledge and understanding of the Universe has been revolutionised since the overturning of the geocentric model of Earth at its centre.

FUTURE EXPLORATION AND INVESTIGATION

Advances in technology have seen the human race break free of the bounds of Earth's gravity and send more than 550 astronauts and cosmonauts as well as large numbers of machines into space for the first time. Whilst some probes have physically travelled to key targets around the Solar System, much of the greatest successes and key advances in knowledge have come from observations made by telescopes and instruments either on the ground such as the VLA and GTC or in space, including the Spitzer, Gaia, Chandra and, of course, the Hubble. Space tourism, the prospect of manned travel to the Red Planet or a human return to the Moon often dominate media coverage, but astronomers,

astrophysicists and cosmologists are often more concerned with the work of probes, observatories, telescope arrays and experiments which may refine our knowledge and help answer the large number of questions and mysteries that remain about space from the nature of dark energy and dark matter to understanding black holes, the origins of the Universe and the continuing search for extra-terrestrial life.

Launching in 2021, the James Webb Space Telescope (JWST) is the $9.7 billion successor to the Hubble and will be able to study objects in the infrared portion of the electromagnetic spectrum 400 times fainter than rival telescopes. The next wave of planetary probes include

BepiColombo, which is expected to arrive at Mercury in 2025 and Venera-D, Russia's reboot of its path-finding Venus explorers. The moons of the gas giants, Jupiter and Saturn are targets for several 2020s spacecraft including NASA's Europa Clipper and the European Space Agency's JUICE (JUpiter ICy moons Explorer). Back on Earth, the world's largest optical telescope, the Extremely Large Telescope (ELT), and largest radio telescope, the Square Kilometre Array (SKA), are both expected to be at work by the end of the 2020s and may transform what we have already learned about the Universe.

ABOVE LEFT: A prototype of the high-gain antenna on NASA's Europa Clipper spacecraft undergoes testing in the Experimental Test Range at NASA's Langley Research Center in Hampton, Virginia.

BELOW: A technician stands in
front of a selection of hexagonal
mirrors from the James Webb Space
Telescope (JWST).

189

ABOVE: Six James Webb Space Telescope mirrors are removed after testing at Marshall Space Flight Center in Huntsville, Alabama.

BELOW: This artist's rendering depicts NASA's Europa Clipper spacecraft, reaching its target – Europa – during the late 2020s or 2030s. The spacecraft is designed to a perform a detailed investigation of the Jovian moon's surface and subsurface through repeated fly-bys.

CREDITS